COMPOSITES AND NANOCOMPOSITES

Advances in Materials Science
Volume 4

COMPOSITES AND NANOCOMPOSITES

Edited by
**A. K. Haghi, PhD, Oluwatobi Samuel Oluwafemi, PhD,
Hanna J. Maria, and Josmin P. Jose**

Apple Academic Press

TORONTO   NEW JERSEY

© 2013 by
Apple Academic Press Inc.
3333 Mistwell Crescent
Oakville, ON L6L 0A2
Canada

Apple Academic Press Inc.
9 Spinnaker Way, Waretown, NJ 08758
USA

First issued in paperback 2021

Exclusive worldwide distribution by CRC Press, a Taylor & Francis Group

ISBN 13: 978-1-77463-260-4 (pbk)
ISBN 13: 978-1-926895-28-4 (hbk)

Library of Congress Control Number: 2012951937

Library and Archives Canada Cataloguing in Publication

Composites and nanocomposites/edited by A.K. Haghi ... [et al.].

(Advances in materials science; v. 4)
Based on papers presented at the First International Conference on Composites and Nanocomposites (ICNC 2011), Jan. 7, 8, and 9, at Mahatma Gandhi University, Kottayam, Kerala, India.

Includes bibliographical references and index.
ISBN 978-1-926895-28-4
1. Composite materials–Congresses. 2. Nanocomposites (Materials)–Congresses. 3. Materials science--Congresses. I. Haghi, A. K II. International Conference on Composites and Nanocomposites (1st: 2011 : Kottayam, India) III. Series: Advances in materials science (Apple Academic Press) ; v. 4

TA418.9.C6C64 2013   620.1'18   C2012-906416-5

Apple Academic Press also publishes its books in a variety of electronic formats. Some content that appears in print may not be available in electronic format. For information about Apple Academic Press products, visit our website at **www.appleacademicpress.com**

About the Editors

A. K. Haghi, PhD

A. K. Haghi, PhD, is the author and editor of 65 books as well as 1000 published papers in various journals and conference proceedings. He has served as a professor at several universities and is a faculty member at the University of Guilan (Iran) as well as a member of the Canadian Research and Development Center of Sciences and Cultures (CRDCSC), Montreal, Quebec, Canada.

Oluwatobi Samuel Oluwafemi, PhD

Oluwatobi Samuel Oluwafemi, PhD, is a Senior Lecturer at the Department of Chemistry and Chemical Technology, Walter Sisulu University, Mthatha Campus, Eastern Cape, South Africa. He has published many papers in internationally reviewed journals and has presented at several professional meetings. He is a fellow of many professional bodies, a reviewer for many international journals, and has received many awards for his work in materials research.

Hanna J. Maria

Hanna J. Maria is a research scholar in the School of Chemical Sciences at Mahatma Gandhi University, Kottayam, Kerala, India. Her research interests include polymer blends ,compatibilization of polymer blends,polymer blend nanocomposites etc.

Josmin P. Jose

Josmin P. Jose is currently pursuing her PhD in nanocomposites at Mahatma Gandhi University, Kottayam, Kerala, India. She has presented papers at several conferences. Her area of interest is polymer nanocomposites for dielectric applications.

Advances in Materials Science

Series Editors-in-Chief

Sabu Thomas, PhD

Dr. Sabu Thomas is the Director of the School of Chemical Sciences, Mahatma Gandhi University, Kottayam, India. He is also a full professor of polymer science and engineering and Director of the Centre for nanoscience and nanotechnology of the same university. He is a fellow of many professional bodies. Professor Thomas has authored or co-authored many papers in international peer-reviewed journals in the area of polymer processing. He has organized several international conferences and has more than 420 publications, 11 books and two patents to his credit. He has been involved in a number of books both as author and editor. He is a reviewer to many international journals and has received many awards for his excellent work in polymer processing. His h Index is 42. Professor Thomas is listed as the 5th position in the list of Most Productive Researchers in India, in 2008.

Mathew Sebastian, MD

Dr. Mathew Sebastian has a degree in surgery (1976) with specialization in Ayurveda. He holds several diplomas in acupuncture, neural therapy (pain therapy), manual therapy and vascular diseases. He was a missionary doctor in Mugana Hospital, Bukoba in Tansania, Africa (1976-1978) and underwent surgical training in different hospitals in Austria, Germany, and India for more than 10 years. Since 2000 he is the doctor in charge of the Ayurveda and Vein Clinic in Klagenfurt, Austria. At present he is a Consultant Surgeon at Privatclinic Maria Hilf, Klagenfurt. He is a member of the scientific advisory committee of the European Academy for Ayurveda, Birstein, Germany, and the TAM advisory committee (Traditional Asian Medicine, Sector Ayurveda) of the Austrian Ministry for Health, Vienna. He conducted an International Ayurveda Congress in Klagenfurt, Austria, in 2010. He has several publications to his name.

Anne George, MD

Anne George, MD, is the Director of the Institute for Holistic Medical Sciences, Kottayam, Kerala, India. She did her MBBS (Bachelor of Medicine, Bachelor of Surgery) at Trivandrum Medical College, University of Kerala, India. She acquired a DGO (Diploma in Obstetrics and Gynaecology) from the University of Vienna, Austria; Diploma Acupuncture from the University of Vienna; and an MD from Kottayam Medical College, Mahatma Gandhi University, Kerala, India. She has organized several international conferences, is a fellow of the American Medical Society, and is a member of many international organizations. She has five publications to her name and has presented 25 papers.

Dr. Yang Weimin

Dr. Yang Weimin is the Taishan Scholar Professor of Quingdao University of Science and Technology in China. He is a full professor at the Beijing University of Chemical Technology and a fellow of many professional organizations. Professor Weimin has authored many papers in international peer-reviewed journals in the area of polymer processing. He has been contributed to a number of books as author and editor and acts as a reviewer to many international journals. In addition, he is a consultant to many polymer equipment manufacturers. He has also received numerous award for his work in polymer processing.

Contents

List of Contributors

Z. Ahmad
Faculty of Civil Engineering, Universiti Teknologi Mara, Malaysia.

M. P. Ansell
Universiti of Bath, United Kingdom.

Tessy Theres Baby
Alternative Energy and Nanotechnology Laboratory (AENL), Nano Functional Materials Technology Centre (NFMTC), Department of Physics, Indian Institute of Technology Madras, Chennai, India–600036.

D. P. Bisen
Pt. Ravishankar Shulka University Raipur (Chhattisgarh).

Nameeta Bramhe
Pt. Ravishankar Shulka University Raipur (Chhattisgarh).

B. Cerroni
Dipartimento di Scienze e Tecnologie Chimiche, and CRS SOFT CNR-INFM, Universita` di Roma Tor Vergata, 000133 Roma, Italy.

E. Chiessi
Dipartimento di Scienze e Tecnologie Chimiche, and CRS SOFT CNR-INFM, Universita` di Roma Tor Vergata, 000133 Roma, Italy.

Sangram K. Das
Institute of Materials Science, Planetarium Building, Bhubaneswar–751013, Orissa, India.

J. C. Diez
ICMA (CSIC-Universidad de Zaragoza), Dpto. de Ciencia de Materiales, C/María de Luna 3, Zaragoza–50018 (Spain).

L. C. Estepa
ICMA (CSIC-Universidad de Zaragoza), Dpto. de Ciencia de Materiales, C/María de Luna 3, Zaragoza–50018 (Spain).

R. Fink
Physikalische Chemie II, Friedrich-Alexander Universität Erlangen- Nürnberg, Egerlandstrasse 3, D-91058 Erlangen, Germany.

S. V. Ghugare
Dipartimento di Scienze e Tecnologie Chimiche, and CRS SOFT CNR-INFM, Universita` di Roma Tor Vergata, 000133 Roma, Italy.

Dr. P. Govindasamy
Director, Erode Builder Educational Trust's Group of Institutions, Kangayam, Tamilnadu, India.

Nursia Hassan
Department of Bioprocess Engineering, Faculty of Chemical Engineering, Universiti Teknologi Malaysia, 81310 Skudai, Johor, Malaysia.

Jiaye Huang
School of Materials Science and Engineering, Beihang University, 37 Xueyuan Road, Haidian District, Beijing 100191, China.

Ani Idris
Department of Bioprocess Engineering, Faculty of Chemical Engineering, Universiti Teknologi Malaysia, 81310 Skudai, Johor, Malaysia.

G. Paruthimal Kalaignan
Advanced Nanocomposite Coatings Laboratory, Department of Industrial Chemistry, Alagappa University, Karaikudi–630 003, Tamilnadu, India.

K. B. Katnam
Irish Centre for Composites Research (IComp), Materials and Surface Science Institute (MSSI), University of Limerick, Ireland.

P. P. Lizymol
Dental Products Laboratory, Biomedical Technology Wing Sree Chitra Tirunal Institute for Medical Sciences and Technology Poojappura, Thiruvananthapuram 695012.

M. A. Madre
ICMA (CSIC-Universidad de Zaragoza), Dpto. de Ciencia de Materiales, C/María de Luna 3, Zaragoza–50018 (Spain).

Siby Mathew
School of Pure & Applied Physics, Mahatma Gandhi University, Kottayam–686560, India.

C. Meneghini
University of Rome ROMA TRE, physics department *via* della Vasca Navale 84, I–00146 Rome, Italy.

Viswaranjan Mohanta
Institute of Materials Science, Planetarium Building, Bhubaneswar–751013, Orissa, India.

Audrey-Flore Ngomsik
Green chemistry and Processes for sustainable development, Université de Reims Champagne Ardenne, Bâtiment 13 Recherche, 57 bis rue Pierre Taittinger 51096 Reims Cedex France.

G. Paradossi
Dipartimento di Scienze e Tecnologie Chimiche, and CRS SOFT CNR-INFM, Universita` di Roma Tor Vergata, 000133 Roma, Italy.

Sangram K. Pradhan
Institute of Materials Science, Planetarium Building, Bhubaneswar–751013, Orissa, India.

Sundara Ramaprabhu
Alternative Energy and Nanotechnology Laboratory (AENL), Nano Functional Materials Technology Centre (NFMTC), Department of Physics, Indian Institute of Technology Madras, Chennai, India–600036.

B. Ranjith
Advanced Nanocomposite Coatings Laboratory, Department of Industrial Chemistry, Alagappa University, Karaikudi–630 003, Tamilnadu, India.

Sh. Rasekh
ICMA (CSIC-Universidad de Zaragoza), Dpto. de Ciencia de Materiales, C/María de Luna 3, Zaragoza–50018 (Spain).

G. Ravichandran
Department of I & P Engineering, the National Institute of Engineering, Mysore–570008, India.

C. S. Robinson
Bhilai Institute of Technology Bhilai House Durg (Chhattisgarh).

Binod K. Roul
Institute of Materials Science, Planetarium Building, Bhubaneswar–751013, Orissa, India.

Prajna P. Rout
Institute of Materials Science, Planetarium Building, Bhubaneswar–751013, Orissa, India.

K. Sakthivadivel
Assistant Professor, Department of Mechanical Engineering, Faculty of Engineering, Erode Builder Educational Trust's Group of Institutions, Kangayam, Tamilnadu, India.

Wei Sha
School of Planning, Architecture and Civil Engineering, Queen's University Belfast, Belfast BT7 1NN, United Kingdom.

D. Smedley
Rotafix Ltd, United Kingdom.

A. Sotelo
1ICMA (CSIC-Universidad de Zaragoza), Dpto. de Ciencia de Materiales, C/María de Luna 3, Zaragoza–50018 (Spain).

B. Suresha
Department of Mechanical Engineering, the National Institute of Engineering, Mysore–570008, India.

P. Md Tahir
Institute of Tropical Forest Product, Universiti Putra Malaysia.

Raunak Kumar Tamrakar
Bhilai Institute of Technology Bhilai House Durg (Chhattisgarh).

Sunil Thomas
School of Pure and Applied Physics, Mahatma Gandhi University, Kottayam–686560, India.

M. A. Torres
Universidad de Zaragoza, Dpto. de Ingeniería de Diseño y Fabricación, C/María de Luna 3, Zaragoza–50018 (Spain).

N. V. Unnikrishnan
School of Pure and Applied Physics, Mahatma Gandhi University, Kottayam–686560, India.

Y. S. Varadarajan
Department of I & P Engineering, the National Institute of Engineering, Mysore–570008, India.

Prathibha Vasudevan
School of Pure and Applied Physics, Mahatma Gandhi University, Kottayam–686560, India.

Guoqing Wu
School of Materials Science and Engineering, Beihang University, 37 Xueyuan Road, Haidian District, Beijing–100191, China.

T. M. Young
Irish Centre for Composites Research (IComp), Materials and Surface Science Institute (MSSI), University of Limerick, Ireland.

List of Abbreviations

2D	Two dimensional
AF	Aramid fibers
AFM	Atomic force microscope
ANOVA	Analysis of variance
BET	Brunauer-Emmett-Teller
Bis-GMA	Bisphenol A- Glycidyl methacrylate
CdS	Cadmium sulfide
CF	Carbon fibers
CFRP	Carbon fiber reinforced polymer
CLSM	Confocal laser scanning microscopy
CQ	Camphor quinone
CS	Compressive strength
CTBN	Carboxyl-terminated butadiene acrylonitrile
DAFS	Diffraction anomalous fine structure
DC	Direct current
DCB	Double cantilever beam
DGEBA	Diglycidylether of bisphenol-A
DMAP	4-(N, N-Dimethylamino) pyridine
DMSO	Dimethyl sulfoxide
DMTA	Dynamic mechanical thermal analysis
DOX	Doxorubicin
DPC	Diphenylcarbazide
DS	Degree of substitution
DSC	Differential scanning calorimetry
DTS	Diametral tensile strength
EDX	Energy dispersive spectroscopy
ED-XAFS	Energy dispersive X-ray absorption fine structure
EXAFS	Extended X-ray absorption fine structure
FBS	Fetal bovine serum
FESEM	Field emission scanning electron microscopy
FITC	Fluorescein isothiocynate isomer
FM	Ferromagnetic
FM	Flexural modulus
FN	Fowler–Nordheim
FPDs	Flat panel displays
FRP	Fiber reinforced plastic
FS	Flexural strength

FT	Field-Oriented texture
FTIR	Fourier transform infra red
G-E	Glass-epoxy
GF	Glass fibers
GFRP	Glass fiber reinforced plastic
GMA	Glycidylmethacrylate
GO	Graphite oxide
HD	High definition
ICDD	International centre for diffraction data
LCA	Linear combination analysis
LFZ	Laser floating zone
ME	Magnetoelectric
MEK-U	Methyl-ethyl-ketone wipe and ultrasonic
MTT	Thiazolyl blue tetrazolium bromide
NEXAFS	Near-edge X-ray absorption fine structure
NIPAAm	N-isopropyl acrylamide
NMP	n-methyl-2-pyrrolidinone
NPs	Nanoparticles
OM	Optical microscopy
PBS	Phosphate buffered saline
PC	Pulse current
PF	Power factor
PMA	Perpendicular magnetic anisotropy
PMT	Photomultiplier tube
PNIPAAm	Poly(N-isopropyl acrylamide
PPD	1-phenyl-1,2-propanedione
PTVs	Projection televisions
PVA	Polyvinyl alcohol
PVP	Polyvinylpyrrolidone
RTGG	Reactive templated grain growth
SCE	Saturated calomel electrode
SEI	Secondary electron image
SEM	Scanning electron microscopy
STXM	Scanning transmission X-ray microscopy
SW-XAFS	Standing waves-XAFS
TEG	Thermally exfoliated graphene
TEGDMA	Triethylene glycol dimethacrylate
TEM	Transmission electron microscopy
TEP	Thermoelectric power
TEY	Total electron yield
TIOT	Tetraisopropylorthotitanante
TL	Thermoluminescence

TR-XAFS	Time resolved XAFS
UC	Upconversion
VHN	Vickers hardness number
VPTT	Volume phase transition temperature
XAFS	X-ray absorption fine structure
XEOL	X-ray excited optical luminescence
XPS	X-ray photoelectron spectroscopy
XRD	X-ray diffraction
XSW	X-ray standing wave
XWG	X-ray waveguides

Preface

Composites are a class of material that receives much attention not only because it is on the cutting edge of active materials research fields due to the appearance of many new types of composites, e.g., nanocomposites and bio-medical composites, but also because there is a great deal of promise for their potential applications in various industries, ranging from aerospace to construction. Technological advances in the composite field are included in the equipment surrounding us daily; our lives are becoming safer, hand in hand with economical and ecological advantages. Researchers are finding ways to improve other qualities of composites so that they may be strong, lightweight, long-lived, and inexpensive to produce. Preparation of nanocomposites also pose very real processing challenges. The list of questions about the fabrication, characterization, and use of nanocomposites is long, despite the massive financial and intellectual investment. The magnitude of the effects these small particles impart to the bulk properties of a composite is great enough that the research in this scientific area will continue to grow in importance.

ICNC 2011 the First International Conference on Composites and Nanocomposites (January 7, 8 and 9) took place at Mahatma Gandhi University, Kottayam, Kerala, India. It was jointly organized by Beijing University of Chemical Technology, China; Venen - Klinik, Austria, and the Institute for Macromolecular Science and Engineering. The conference was designed to be the perfect opportunity for international researchers interested in composites and nanocomposites to meet, present, and discuss issues related to current developments in the field of composite and nanocomposites. The goal of the conference was to promote interdisciplinary research on processing, morphology, structure, properties, and applications of macro, micro, and nanoparticles, their composites and their applications in medicine, automotive, civil, chemical, aerospace, computer, and marine engineering. The 3-day conference included discussion on recent advances, difficulties, and breakthroughs in the field of composites and nanocomposites. Synthesis and characterization of macro, micro and nanoparticles, magnetic and polymeric macro, micro and nanocomposites, bionanocomposites, rheology of macro, micro and nanocomposites, nanoparticles for drug delivery system, fully green macro, micro and nanocomposites, processing of macro, micro and nanocomposites, applications of macro, micro and nanocomposites, etc., were a few among the wide area of topics covered at the conference. The conference included keynote addresses, a number of plenary sessions, invited talks, and contributed lectures, focusing on specific tenets of composites and nanocomposites. Additionally, there was a poster session with more than 40 posters to encourage growing scientists and researchers in this field. The conference had over 200 delegates from all across the world, with good representation from France, Australia, USA, Netherlands, Germany, Iran, Sweden, Spain, and Libya, as well as from a substantial number of eminent Indian scientists.

This book, *Composites and Nanocomposites*, collects together a selection of 15 papers presented during the conference. The papers include a wide variety of interdisciplinary topics in composites and nanocomposites. This body of papers from leading experts in this field, both fundamental and applied, constitutes a critical mass that will provoke further discussion in this broad interdisciplinary area. We appreciate the efforts and enthusiasm of the contributing authors for this topic and acknowledge those who were prepared to contribute but were unable to do so at this time. We trust that this volume will stimulate new ideas, methods, and applications in ongoing advances in this growing area of strong international interest.

We would like to thank all who kindly contributed their papers for this book and the editors of Apple Academic Press for their kind help and cooperation. We are also indebted to the Apple Academic Press editorial office and the publishing and production teams for their assistance in preparation and publication of this issue.

**— A. K. Haghi, PhD, Oluwantobi Samuel Oluwafemi, PhD,
Josmin P. Jose, and Hanna J. Maria**

1 Finite Element Modeling of Properties Influence of Particle Characteristics in Particle Reinforced Metal Matrix Composite

Jiaye Huang, Guoqing Wu, and Wei Sha

CONTENTS

1.1 INTRODUCTION

The particles adopted are usually non-metallic such as ceramics and graphite, as they tend to have properties by their own. The main factors to be considered when choosing a type of particle should include shape and size, physical properties, mechanical properties, processing, and its compatibility with the matrix. The currently widely used particle reinforcements are SiC, BC, and Al_2O_3, which are effective in increasing composite strength and modulus. However, the drawback is the significant loss of ductility. The ceramic particles are brittle materials. Under stress, the particles themselves, and the particle-particle and the particle-matrix interfaces can all fracture, leading to the composite failure.

A new Mg–Li matrix composite with 5 wt% YAl_2 particulates was developed by stir casting technique [1]. Its microstructures and properties were investigated systematically. The results show that the YAl_2 intermetallics particles distribute in Mg–Li matrix homogeneously, and a good YAl_{2p}/Mg–Li interface is developed wherein there are no reaction products or obvious elements diffusing. The composite has a higher tensile strength compared with matrix alloy, whose good ductility is kept. It has been found that the intermetallic YAl_2 when used as particle reinforcement can be beneficial to the composite ductility while at the same time effectively increasing composite strength. Microscopy showed good interface connection between the reinforcing particle and the matrix, without voids, interface fracture, interface reaction, and amorphous layer formation. These all contribute to the better composite properties.

In order to identify the mechanisms of particle reinforcement in metal matrix composites, we need to consider the following two aspects when setting up models:

(1) The distribution and shape of the particles in the model must be representative of the real composite material. The size of the model should be as large as possible to reduce boundary effects.

(2) A true reflection of the particle reinforcing effect, especially in terms of the good interface between particles and matrix as observed experimentally. The model needs to describe and simulate such interface correctly.

In the past, many researchers have designed axisymmetrical unit cell models containing one reinforcing particle [2], spheroidal unit cell model incorporating interactions between particles [3], cubic model of randomly distributed spherical particles [4], and cell model containing interface [5] using the serial sectioning method [6, 7]. These models can simulate to a certain extent the tensile and fracture behavior of particle reinforced metal matrix composite materials. Using these models, the effects of reinforcing particle volume fraction, shape, size, and interface on the mechanical properties of the composite materials have been analyzed. Some results have emerged. For example, it has been found that smaller reinforcing particles are more effective in improving properties of composite materials and minimizing failure of the materials [8]. Circular and smooth reinforcing particles reduce stress concentration in materials and thus minimize fracture at the interface [9]. Also, it has been found that uniform particle distribution is beneficial to homogeneous stress and strain distribution during the material deformation process, thus avoiding local stress concentration [10, 11].

All the existing models, while solving some problems, have their limitations. Examples of these are the assumption of periodicity of particle distribution, the assumption of spherical particles, the difficulty in software manipulation, and the lack of universal applicability of the software. In addition, there is some way to achieving a comprehensive theoretical framework, capable of evaluating all relevant factors.

Therefore, constructing more realistic microscale models of particle reinforced metal matrix composites, representing real material conditions and with universal software applicability, is an important and immediate task. Such models will help exploring the effects of various factors on the composite material properties, and directing the future research and development of new composite material systems.

An interface transition region can be formed through the following ways:

(1) Chemical reaction at the interface between the reinforcing body and the matrix

(2) Coating on the reinforcing body

(3) Diffusion of elements across the interface

(4) Even the mechanical interface, where the reinforcing body and the matrix are mounted mechanically at the interface, will have finite thickness with varying bonding strength across the interface thickness.

Considering the effect of particle shape, distribution, volume fraction, and interface, this chapter proposes a two-dimensional (2D) model of randomly distributed spheroidal particles [12] coupled with an axisymmetric unit cell model containing one reinforcing particle. The aim is to develop a platform for further coupled computation based on micro as well as macro models for simulating micro phenomena as well as experimental validation of composite materials.

It is well known that the physical and mechanical properties of metal matrix composites are strongly dependent on the characteristics, size, and number distribution of the reinforcing particles, but we will now explain why the interface strength between the matrix and the reinforcing particles is as important. One of the main reasons of the significantly increased strength of a metal matrix composite over the metal matrix is that the reinforcing phase can take part of the load on the composite. The load transfer from the matrix to the reinforcing body is through the interface between them, and therefore the strengthening mechanisms are strongly related to the strength of the interface. A good interface with good connection and high interface strength can transfer the load effectively, and help increase the composite strength. Conversely, a poorly connected interface will not be ideal for strengthening.

Therefore, the interface quality determines the efficiency of the load transfer from the matrix to the reinforcing body. There are several types of interface connection in composite materials:

(1) Simple connection interface, without solution, diffusion, reaction, and having good wettability. This is a clean and tight-bonding interface. Semi-coherent atom matching interface belongs to this type of interface.

(2) Solution or diffusion interface: As the name implies, there is interdiffusion between atoms in the reinforcing particles and the matrix in the interface region.

(3) Reactive interface, due to the formation of new chemical compound(s) through chemical reactions at the interface.

(4) Mechanical interface: The reinforcing body and the matrix are mounted mechanically at the interface. Rougher interfaces are beneficial to such mechanically connected interface by making the bonding stronger.

If the interface strength is weak, it will directly affect the strength of the composite system. Under stress conditions, defects usually initiate at the weak interface. Part of this chapter will describe the unit cell model containing one interface layer and examine the effect of the elastic modulus and the yield strength of this single layer on the stress and strain distributions in the composite material. Finally, the discussion will be extended to interfaces with several transition layers forming a gradient of properties.

The characteristics of particle reinforcement have significant impact on the performance improvement of particle-reinforced composites, which include particle size distribution, shape, volume fraction, and the nature of the interface. A model of a 2D randomly distributed spheroidal particles coupled with an axisymmetric unit cell model containing one reinforcing particle with a transition interface was proposed. Macroscopic mechanical properties were simulated with the 2D randomly distributed spheroidal particles model and the effects of interface characteristics were discussed in the single reinforcing particle axisymmetric unit cell micromodel. This micromodel is developed considering the supposed impact of the interface compatibility between the reinforcements and the matrix. The influence of interfaces on the composite modulus and the stress-strain distribution was discussed with this model. It was shown that composites with transition interfaces could bear higher stresses than those with simple zero thickness interfaces.

1.2 THE PROPOSAL OF COUPLED MODEL OF PARTICLE REINFORCED METAL MATRIX COMPOSITES

The models in the past were mostly unit cell models having various shapes [4]. Randomly distributed spherical particles were usually considered. Upon applying external stress, the internal stress distribution can be calculated. In reality, however, the particles in a composite material would not be periodically distributed as often assumed in unit cell models. In addition, affected by processing, the particle shape is not necessarily spherical, but is more likely spheroidal. In the present program of works, we have established a randomly distributed spheroidal particle model (Figure 1) [12]. Based on literature and experimental data and cross validation, the effects of particle material parameters, geometrical parameters, and volume fractions on the composite tensile properties are discussed [12].

During experiments using the intermetallic YAl_2 as particle reinforcement, it was found that the good interface connection between this kind of reinforcement and the matrix is possibly the reason for maintaining a good level of ductility while increasing the modulus and strength. Based on this finding, we have set up a unit cell model including interface layers, in order to investigate the effect of the interface elastic modulus and the interface strength on the composite material. The objective was to set up a model describing the interface transition layer. Therefore, combining the models described [12] and in the following part of this study, we will be able to obtain a coupled model describing the entire particle reinforced metal matrix composite structure (Figure 2). The key concept in the model focused next in this study is the expression of the interface transition layer, enabling simulating the effects of gradual, that is, not sharp interface between the particle and the matrix.

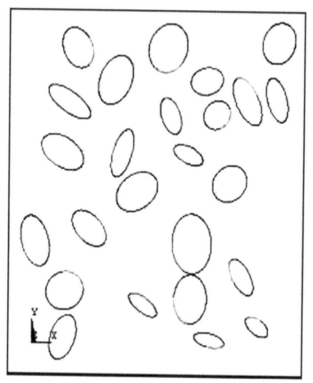

FIGURE 1   The model generated randomly distributed spheroidal particles.

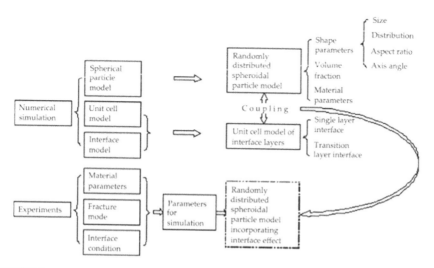

FIGURE 2   The coupled model of particle reinforced metal matrix composites.

The macro and micro models are based on the same reinforcing particles and the matrix material, but with different shape and distribution. The material parameters are given [12]. When not considering fracture, the main material phenomena under external stress are elastic and plastic deformation, and thus the elastic modulus and the yield strength are the main parameters. We have described the model used for simulation [13].

1.3 MODEL WITH ONE INTERFACE LAYER AND THE EFFECTS OF ITS MODULUS AND STRENGTH

This study will start the simulation and discussion of the effects of the interface layers on composite properties, and build a foundation for further investigation of such effects. The aim is to set up a transition interface and evaluate the necessity and effectiveness of using the transition interface in modelling. To achieve this, we first use a unit cell model, apply a uniaxial stress, and calculate the stress and strain in the composite for varying elastic modulus and yield strength of one transition interface layer.

As there is no experimental data available for the material parameters of the interface layer, we make an assumption that this layer has properties in between the matrix and reinforcing body, when the interface bonding is good. With this assumption, the base material parameters adopted in modelling are listed in Table 1, where the interface parameters are around averages of data from the matrix and the reinforcement.

TABLE 1   Materials parameters of matrix, reinforcing body, and the interface.

	Elastic modulus (GPa)	Poisson's ratio	Yield strength (MPa)	Strain hardening rate (MPa)
Matrix	42	0.33	94	200
Interface	100	0.27	1000	160
Reinforcing body	158	0.205	1800	120

1.3.1   Effect of Interface Elastic Modulus on Composite Properties

The calculations in this study use the base material parameters as shown in Table 1, except that the interface elastic modulus is varied above and below the base value of 100 GPa. With varying elastic modulus of the interface layer, under a fixed load of 100 MPa that is slightly larger than the matrix yield strength, the maximum stress of the interface and the maximum stress within the composite system are given in Table 2. It can be seen that when the interface modulus is lower than 100 GPa, the interface maximum stress does not change a great deal, and remain slightly lower than the maximum stress within the composite system. However, the stress taken by the interface is always greater than the average applied stress, that is, load. Therefore, when the interface has a high yield strength, of 1,000 MPa, it having low elastic modulus does not affect the composite properties much.

TABLE 2   Maximum stress in the interface and in the composite system for interface of different elastic modulus and composite under 100 MP loading.

Interface elastic modulus (GPa)	Interface maximum stress (MPa)	Composite maximum stress (MPa)
10	116	129
100	119	140
150	137	137
300	238	238
500	360	360
1000	589	589

When the interface modulus is greater than 150 GPa, that is, greater than the reinforcing body's modulus, the maximum stress in the entire composite system occurs at the interface. Figure 3 shows the internal stress distribution in the composite system when the interface elastic modulus is 1,000 GPa. It can be seen that the interface undertakes the largest stress. This is detrimental to the composite when the interface strength is not very high. Therefore, the modelling shows that the elastic modulus of the interface is significant to the performance of the composite.

589 MPa

31 MPa

FIGURE 3   Stress distribution in the composite for an interface elastic modulus of 1,000 GPa and external load of 100 MPa.

Next, we change the loading to 80 MPa, that is, just below the matrix yield strength, and 120 MPa, that is, just above the matrix yield strength, and examine the trend of change in the total strain of the composite system for different interface elastic modulus. The results are given in Table 3. With the external loading of 80 MPa, when the interface modulus is not very small (greater than 1 GPa), the effect of the interface modulus on the tensile property of the composite material is small. With a 100 fold change of the modulus, from 1 to 100 GPa, the strain changes by about 15%. When the modulus is reduced to 0.001 GPa from 1 GPa, however, the strain increases fifty times. When the loading is 120 MPa and the modulus of the interface is above 0.1 GPa, the strain changes within 10%. When the modulus is reduced to 0.001 GPa, the strain changes by about six times.

TABLE 3   The total strain for different interface modulus and composite under loading of 80 MPa and 120 MPa.

Interface elastic modulus (GPa)	80 MPa loading	120 MPa loading
0.001	7.93%	26.2%
0.01	1.61%	10.8%
0.1	0.287%	4.66%
1	0.159%	4.24%
10	0.141%	4.21%
100	0.139%	4.21%

Summarizing the calculation results, the elastic modulus of the interface layer influences the stress distribution in the composite material and its total strain. When the interface elastic modulus is greater than the reinforcement elastic modulus, the interface layer will attract the largest stress concentration. On the other hand, when the interface elastic modulus is lower than the matrix elastic modulus by more than two orders of magnitude, the interface is very easy to elastically deform, leading to very large strains of the composite system.

1.3.2   Effect of Interface Yield Strength on Composite Properties

We have already seen that too large or too small elastic modulus of the interface layer is detrimental to the mechanical performance of the composite system. In this study, when discussing the effect of the interface yield strength, in order to minimize the influence of its elastic modulus, it is fixed at 100 GPa, that is, midway between the modulus values of the matrix and the reinforcing materials. Table 4 gives the maximum stress in the interface layer for different interface yield strength and the externally applied load of 10 and 100 MPa.

TABLE 4 Maximum stress in the interface for different interface yield strength and composite under loading of 10 MPa and 100 MPa.

Interface yield strength (MPa)	10 MPa loading	100 MPa loading
10	11.9	100
100	11.9	119
500	11.9	119
1000	11.9	119

From Table 4, when the interface yield strength is increased from 10 to 1,000 MPa, the maximum interface stress does not change significantly. If the interface yield strength is low, the stress at the interface is higher than its yield strength, and it will become the weakest region in the composite system. With increasing yield strength of the interface, it cannot take much more stress and so cannot effectively relieve the stress concentration inside the material. Therefore, when the external load is not very large, increasing interface yield strength does not change significantly the internal stress distribution.

For a high interface yield strength fixed at 1,000 MPa, to ensure no yielding of the interface, the calculated maximum stress of the interface for different loading of 10, 100, and 500 MPa is 11.9, 119, and 775, respectively, the first two results already included in Table 4. Under such conditions with no interface yielding due to the fixed high interface yield strength, with increasing external load, the maximum interface stress increases as well and remain larger than the external stress.

In summary of the calculations in this study, the interface stress concentration is quite stable and is always larger than the externally applied load. The change of the interface yield strength does not change significantly its stress level. Therefore, it can be concluded that as long as the interface yield strength is slightly larger than the externally applied load, the interface layer can effectively withstand the stress concentration in the entire material.

1.4 STRESS ANALYSIS USING A MULTILAYER TRANSITION INTERFACE MODEL

The discussion is all based on the unit cell model containing one interface layer. The conclusion is that both elastic modulus and yield strength of the interface layer do affect the overall composite properties. This proves the earlier discussion about the composite properties being enhanced by good connection between the matrix and reinforcing bodies at their interface.

In reality, the interface between the reinforcing body and the matrix should not be one layer as used. After the interdiffusion and interaction between the reinforcing body and the matrix, the interface properties more likely change gradually. If there is good interface bonding between the reinforcement and the matrix, such interpenetration can be more thorough, forming a thicker transition region with a more gradual change of properties in the transition region. Conversely, if there is bad interface bonding between the reinforcement and the matrix, such interpenetration should be limited, forming a thin transition region, and the property transition would not be as gradual. Under external stress, those interfaces that have large lattice mismatch and where stress concentration is not easily released will become weak regions in the material. Experimental evidence shows that the interface with ceramic reinforcement is usually where fracture starts. In this study, we will use the multilayer interface model to simulate the maximum stress region in the unit cell model [13]. As discussed, this is along the short axis of the particle. We will apply a stress perpendicular to this short axis.

In order to concentrate our investigation on the effect of interface layers on the composite properties, the following assumptions are made, considering that the yield strength of the reinforcing material is far greater than the matrix material. The reinforcement material is regarded as very rigid, and it does not participate in deformation. The interface layer closest to the reinforcement (left-most layer) has the material parameters of the reinforcement. Such parameters decrease, in an evenly stepped fashion, from the layer closest to the reinforcement to the layer closest to the matrix. The rightmost layer is the matrix material. The materials parameters of the different interface layers are shown in Table 5.

TABLE 5   The elastic modulus and the yield strength of the interface layers.

Number of interface layers	Elastic modulus (GPa)	Yield strength (MPa)
1	158	1800
3	158/120/80	1800/1200/600
6	158/140/120/100/80/60	1800/1500/1200/900/600/300

1.4.1 Stress Analysis with Single Interface Layer Model

The model adopting one interface layer means that there is no interpenetration between the reinforcing body and the matrix. The two just contact each other. Applying different stresses, the resulting stress distribution is shown in Figure 4.

From Figure 4, when the externally applied stress is 100 MPa, because the interface layer undertakes some stress concentration, the majority part of the matrix does not yield. The interface layer and the matrix both have elastic deformation, and the maximum stress occurs at the junction between the matrix and the interface, near the loading position (i.e., the top of the modelled area). With increasing external load, after much of the matrix yields, the elastic deformation of the interface layer reduces, and the stress concentration region enlarges. The maximum stress can be more than three times of the loading stress. The part of the matrix near the interface layer has the lowest stress within the entire matrix. Only when the loading reaches 500 MPa does most of the matrix yield. Noting that the matrix yield strength is only 94 MPa, we can see that the high modulus and high strength interface layer shares the majority of the load bearing. Conversely, if the reinforcement strength is low, it is possible to yield far before the average external load reaches its yield strength.

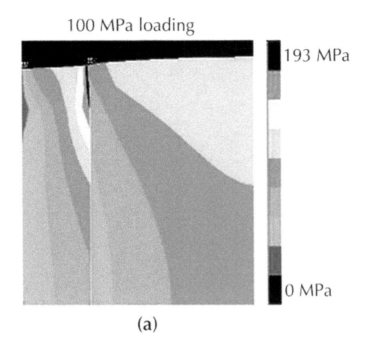

(a)

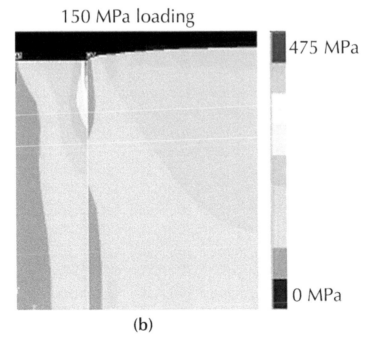

(b)

FIGURE 4   *(Continued)*

200 MPa loading

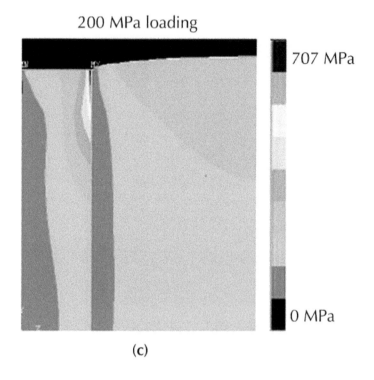

(c)

300 MPa loading

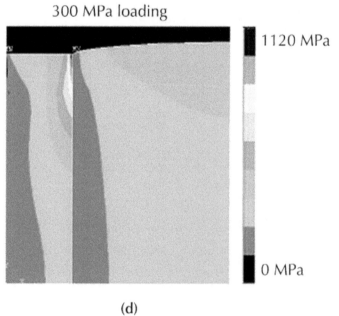

(d)

FIGURE 4   *(Continued)*

500 MPa loading

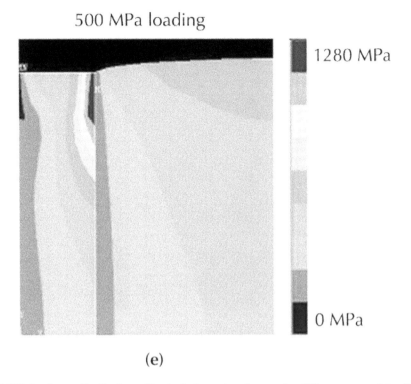

1280 MPa

0 MPa

(e)

FIGURE 4 Stress distribution with single layer interface under different external loading. (a) 100 MPa; (b) 150 MPa; (c) 200 MPa; (d) 300 MPa; (e) 500 MPa.

1.4.2 Stress Analysis with Three Transition Layers

If we increase the number of transition layers from one to three, with elastic modulus 158, 120, and 80 GPa, respectively, and yield strength 1,800, 1,200 and 600 MPa, respectively, the internal stress distribution under different applied stresses is obtained as shown in Figure 5.

From Figure 5, with 100 MPa external loading, because the interface layers undertake some stress, the matrix material does not yield, but elastically deform with the interface layers. The maximum stress occurs in two interface layers, and is lower than in the case of single interface layer, by nearly 40%. Therefore, the continuous interface layers do help withstand the external stress and protect the matrix material. With increasing load, the stress concentration first occurs in the interface layer next to the matrix, but the maximum stress is still much smaller than in the case of single interface layer. At 500 MPa loading, after the interface layer next to the matrix yields, the maximum stress location moves to the second interface layer. Therefore, the interface layers take the stress concentration in turns, effectively as a safeguard of the composite system.

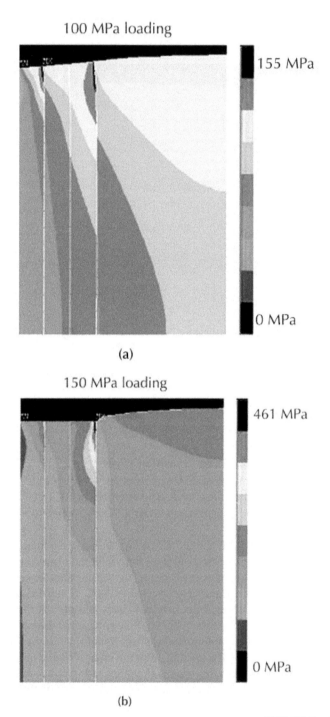

(a)

(b)

FIGURE 5   *(Continued)*

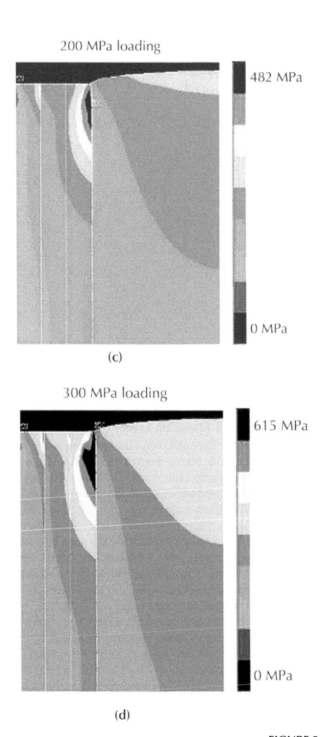

FIGURE 5   *(Continued)*

500 MPa loading

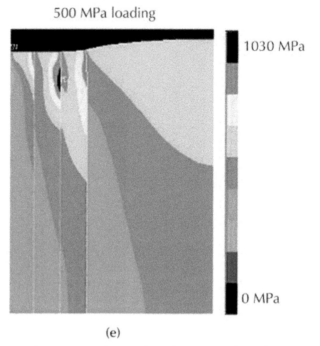

1030 MPa

0 MPa

(e)

FIGURE 5 Stress distribution with three layer interface under different external loading. (a) 100 MPa; (b) 150 MPa; (c) 200 MPa; (d) 300 MPa; and (e) 500 MPa.

1.4.3 Stress Analysis with Six Transition Layers

The complexity of the interface increases with the number of transition layers used. We now increase this to six, with elastic modulus 158, 140, 120, 100, 80, and 60 GPa, respectively, and yield strength 1800, 1500, 1200, 900, 600, and 300 MPa, respectively. Applying different external loading stresses, the internal stress distribution obtained is shown in Figure 6. It can be seen from this figure that when the applied load is 100 MPa, because the interface layers support some stress, the matrix material does not yield, but instead elastically deform with the interface layers. The maximum stress appears uniformly in the matrix and all the interface layers, achieving a more homogeneous distribution of stress. The maximum stress further reduces compared to the three-layer interface. With increasing external load to 150 MPa, the stress concentration first happens in the interface layer next to the matrix, but the maximum stress reduces significantly compared to the three-layer interface under the same load. With 300 MPa applied load, after the first interface layer next to the matrix yields, the maximum stress moves to the second interface layer.

From these results, it can be concluded that with increasing interface layers, under the same load, the internal stress spreads out and becomes more homogeneous, and the maximum stress decreases. With multiple interface layers, after the layer next to the matrix yields, the stress concentration moves to the next layer. This proves that the interface with transition layers can withstand larger stress compared to straight interface between matrix and reinforcing material. A good interface with the matrix may be even more important than the property of the reinforcing material itself.

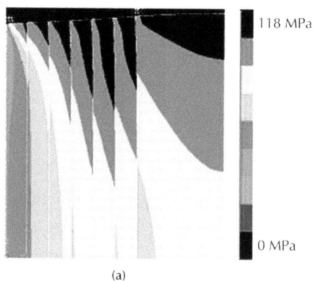

(a)

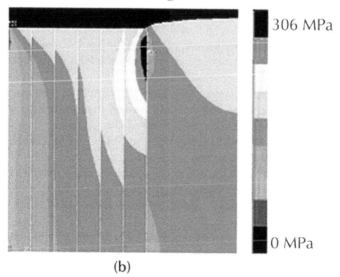

(b)

FIGURE 6   *(Continued)*

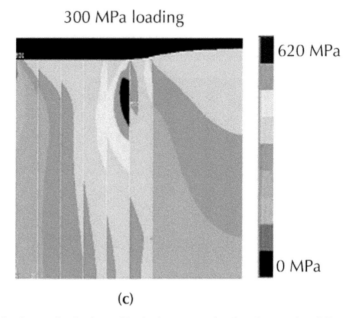

(c)

FIGURE 6 Stress distribution with six layer complex interface under different external loading. (a) 100 MPa; (b) 150 MPa; (c) 300 MPa.

1.4.4   The Effect of the Transition Interface on the Elastic Modulus in a Unit Cell Model

For different conditions of the interface as described, the elastic modulus of the composite material can be calculated, based on the axisymmetric unit cell model. The calculation gives the results of 57.1, 58.2, and 58.8 GPa, for one, three, and six transition layers, respectively. With increasing complexity of the interface transition, that is the increase of the number of transition layers, the elastic modulus of the composite material increases marginally, no more than 3%.

A more complex transition state of the interface means more gradual and smooth transition of the elastic modulus and the yield strength from the reinforcing body and the matrix. In an ideal case, the best transition should be continuous, that is not stepped as the models used here. From the calculation results, it can be expected that such an ideal interface would increase the composite modulus even more. However, the interface only occupies a small fraction of the volume of composite or its model, and its presence should not change fundamentally the elastic modulus of the composite. This is the reason for the relatively small influence as far as the elastic modulus is concerned.

A gradual interface means better compatibility between the reinforcing body and the matrix, with no weakness regions. In such case, the stress concentration region moves in a gradual manner, and will not cause sudden cracking or fracture due to the sudden change of material strength in different regions. If the reinforcement and the matrix do not have such good compatibility, the connection interface will have large and sudden strength change. If there is no transition region at the interface, cracking and failure will happen more easily.

1.5 CONCLUSION

The interface and its properties have significant influence over the yield strength and the elastic modulus of the composite systems. We have concentrated on the interface and simulated the tensile process of the unit cell model containing an interface layer. Simulation results of stress distribution show the region having the highest stress levels. Concentrating on this region, a layered interface model is designed to resemble the gradual change of modulus and strength from the reinforcing particle to the matrix. Physically, such change could be due to, for example, elemental diffusion between the particle and the matrix, causing a gradual decrease of the property levels in the direction from the harder particle to the weaker matrix. In comparison, the past unit cell models assumed simple interfaces with zero thickness. If the transition interface model can represent the actual transitional state of the interface, it can be used to evaluate the effect of the interface on the elastic modulus and the stress and strain conditions of the composite.

   This study has examined the effect of the elastic modulus and the yield strength of the interface layers and obtained the stress distribution within a unit cell model containing interface layers. A model with multiple transition interface layers is established, and the effect of the number of layers on the stress distribution in and around the interface is discussed. The main conclusions are as follows:

(1) The elastic modulus of the interface layer affects the internal stress distribution of the composite material and the total strain. When the elastic modulus of the interface is very high, the interface layer will take most of the stress concentration, and thus risks fracture. When the elastic modulus of the interface is very low, the interface layer can deform greatly, causing large overall strain in the composite system. This would also result in easy cracking at the interface between the matrix and the reinforcing body.

(2) The stress concentration at one layer interface is quite stable, and is always higher than the externally applied load. The variation of the yield strength of the interface does not change significantly its load bearing capability. For the interface layer, its bonding strength controlling fracture is a much more important factor compared to its yield strength.

(3) For interfaces with transition layers, a more gradual and smooth change of material properties within the layers results in spreading and homogenizing of the internal stress. It also reduces the maximum stress existing in the material. After the interface layer next to the matrix yields, the stress concentration moves to the next layer. Thus, an interface with multiple transition layers can effectively increase the load bearing capacity of the composite material, when compared with single, directly bonded interface.

(4) With more gradual and smooth change of material properties in the transition interface layers, the elastic modulus of the composite material increases slowly, as calculated using the unit cell model. This increase, however, is small and is no more than 3% of the initial elastic modulus based on a simple interface.

The problem addressed here is largely artificial. We imagine that the particle-matrix interface can be described by a gradual transition in properties, with no evidence that such large or extended variations might exist. The models are then solved using standard FEM. The methodology is thus similar to many prior works on graded materials, here with an artificial graded interface. Many studies have been published on this broad topic over many years, with more recent work providing insight into actual crack formation.

KEYWORDS

- **Computer simulations**
- **Elasticity**
- **Intermetallics**
- **Mechanical properties**
- **Metal matrix composites**
- **Surfaces and interfaces**

ACKNOWLEDGMENT

This chapter is financially supported by the National Natural Science Foundation of China (50901005) and the Beijing Nova Program (2007B016).

REFERENCES

1. Wang, S. J., Wu, G. Q., Li, R. H., Luo, G. X., and Huang, Z. *Mater. Lett.*, **60**, 1863–1865 (2006).
2. Guild, F. J. and Young, R. J. *J. Mater. Sci.*, **24**, 298–306 (1989).
3. LLorca, J. and Segurado, J. *Mater. Sci. Eng. A*, **365**, 267–274 (2004).
4. Böhm, H. J., Eckschlager, A., and Han, W. *Comput. Mater. Sci.*, **25**, 42–53 (2002).
5. Zhang, P. and Li, F. *Chin. J. Aeronaut.*, **23**, 252–259 (2010).
6. Ganesh, V. V. and Chawla, N. *Mater. Sci. Eng. A*, **391**, 342–353 (2005).
7. Chawla, N., Deng, X., and Schnell, D. R. M. *Mater. Sci. Eng. A*, **426**, 314–322 (2006).
8. Ergun, E. and Aslantas, K. *Mech. Res. Commun.*, **35**, 209–218 (2008).
9. Ramanathan, S., Karthikeyan, R., and Ganasen, G. *Mater. Sci. Eng. A*, **441**, 321–325 (2006).
10. Cheng, N. P. and Zeng, S. M. *Trans. Nonferrous Met. Soc.*, China, **17**, 51–57 (2007).
11. Boselli, J. and Gregson, P. J. *Mater. Sci. Eng. A*, **379**, 72–82 (2004).
12. Huang, J. Y., Ling, Z. H., and Wu, G. Q. *Mater. Sci. Forum*, **650**, 285–289 (2010).
13. Huang, J. Y., Wu, G. Q., and Sha, W. *A transition interface model of particle reinforced metal matrix composite.* In First International Conference on Composites and Nanocomposites (ICNC 2011), Kottayam, Kerala, India, paper 88 (January 7–9, 2011).

2 XAFS Probe in Material Science

C. Meneghini

CONTENTS

2.1 INTRODUCTION

The X-ray absorption fine structure (XAFS) consists of the features appearing in the high energy region above an inner X-ray absorption edge [1-4]: the XAFS features near to the edge (XANES) provide information about the valence state, electronic structure, and local structure topology around the absorber. The analysis of XAFS in the extended region far from the edge (EXAFS) provides details about the local structure around the absorber in terms of pairs and many body distribution functions (Figure 1). One of the major advantage of XAFS technique is the relatively easy data collection and the very broad scope in the material science [6] allowing to study materials in every aggregation state (solid, liquid, gas, and vapor), crystalline or amorphous, from bulk to the nanoscale, thin films, and low dimensional systems, in concentrated as well highly diluted phases. High spatial resolution are reached on the submicrometer scale for detailed mapping [7], specially requested for studying natural samples and cultural heritage investigations [8]. The XAFS is specially suited to study materials under extreme conditions such as high temperature [9], high pressure [10] and high magnetic fields. In addition time resolved XAFS (TR-XAFS) describes the structural and electronic modifications with a resolution reaching the ps scale [11, 12].

The XAFS a basic probe for material science, widely complementary common techniques like x-ray diffraction and laboratory spectroscopy [6]. A brief review of experimental possibilities is presented here with examples specifically related to nanostructured and composite materials.

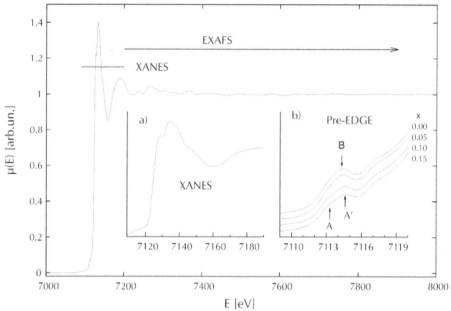

FIGURE 1 The extended and near edge regions of XAFS signal, x-ray absorption signal at Fe K edge Eu doped $Bi_{1-x}Eu_xFeO_3$ multiferroics ceramics. The EXAFS oscillations remain well evident in the high energy region far from the edge. The XANES features are highlighted in the panel (a). The panel (b) shows the pre-EDGE region for different samples as a function of Eu doping: the effects of Eu doping cause the progressive reduction of FeO_6 octhaedra distortions splitting the Fe_{3d} levels T_{2g}-E_{2g} [5] due to crystalline local fields.

2.2 EXPERIMENTAL METHODS AND DATA ANALYSIS

The XAFS data collection consists in measuring a signal proportional to the absorption coefficient μ, as a function of energy, across the absorption edge of a specific element of the sample. Different experimental geometries can be used for that [2], as a function of the sample characteristics: massive bulk samples are measured in transmission geometry, collecting the x-ray incoming flux (I_o), and the flux transmitted (I_t) through a sample of thickness t the absorption signal is obtained as $\mu t = \ln (I_o/I_t)$. Dealing with diluted elements, like biological samples, solutions, thin layers, and quantum dots, it is $\mu t \ll 1$. In that case the fluorescence intensity (I_f) is measured and the total absorption can be calculated as: $\mu \sim (I_f/I_o)$. The I_f can be also measured collecting the total electron yield (TEY). Notice that the surface sensitivity of photoelectron yield, makes TEY geometry exploitable also in case of massive samples (large crystals, for example) in which the fluorescence geometry would be affected by strong self-absorption effects [13, 14]. High quality XAFS data on the last generation synchrotron radiation beam lines reaches highest signal to noise ratio and the widest k-range well above k = 20 Å^1 revealing the finest details of pair and many body distribution functions [1-4].

A peculiar characteristic of XAFS is the directional sensitivity of the absorption process along the X-ray linear polarization vector, this allows to probe electronic and local anisotropy in the samples related to anisotropic physical response of the materials [14-16] (Figure 2). Polarized XAFS is used, for example, to understand the microstructural mechanisms giving rise to the perpendicular magnetic anisotropy (PMA) in thin films and/or quantum dots [16]. A special role in x-ray absorption spectroscopy related techniques is played by the x-ray magnetic circular dichroism (XMCD) which allows to probe element specific magnetic response in the sample, with the unique possibility to distinguish between spin and orbital contributions to the magnetic moment [17].

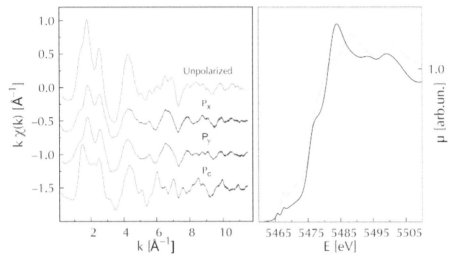

FIGURE 2 Polarized XAFS on 2.8% Cr-doped V_2O_3 single crystal. Low temperature crystallographic phase of V_2O_3 (or $(Cr_{0.014}V_{0.086})_2O_3$) is monoclinic anti-FM insulating (AFI). Left panel compares unpolarized XAFS data, collected on $(Cr_{0.014}V_{0.086})_2O_3$ powders, with the polarized XAFS data collected on a single crystal mounted with different orientations with respect to the x-ray polarization beam. The differences are well evident in the extended region as well in the XANES [14].

The TR-XAFS is exploited to follows the structural and electronic changes related, for example, to the chemical activity of catalysts [6], being relevant to understand the mechanisms related to the catalytic activity and to design more efficient materials. Moderate time resolution ($\Delta t \sim 10^1 s$) is achieved using the quick-XAFS mode [11]. The energy dispersive XAFS (ED-XAFS) geometry ($\Delta t \sim 10^{-3}$-$10^{-5} s$) is used [18] to reach shorter time resolution while pump probe set-up reaches the $\Delta t \sim 10^{-9} s$ [12]. The ED-XAFS geometry has the advantage of having "no moving parts", these results in extreme stability and reproducibility of the data, well suited to emphasize the finest structural and electronic changes as a function of external parameters, like temperature, pressure, and magnetic fields [19].

It is noticeable as the conceptual simplicity of XAFS data collection stimulated exotic and unconventional experimental set-ups: as a matter of fact a number of applications have been proposed such as the x-ray excited optical luminescence (XEOL)

[20], diffraction anomalous fine structure (DAFS) [21], standing waves-XAFS (SW-XAFS)[22, 23], and Refl-EXAFS [24].

From the point of view of data analysis the XAFS signal is usually divided into the near edge (XANES) and the extended (EXAFS) regions (Figure 3): the scattering theory motivate this distinction dividing the full multiple scattering (near-edge) from the scattering (extended) regions [1, 6]. In the EXAFS region the structural signal is connected in a rather simple way to the structure around the absorber and can be interpreted fitting the experimental data to the theoretical curve given by the standard EXFS formula [1-4, 6]:

$$\chi(k) = S_0^2 \sum_j \frac{A(k, R_j) N_j}{k\, R_j^2} \sin\left(2kR_j + \delta(k,\, R_j)\right) e^{-2k^2 \sigma_j^2} e^{-2R_j/\lambda} \tag{1}$$

The Equation (1) describes the local structure around the absorber as a sum of partial contributions (shells) of Gaussian shape, each of one characterized by a multiplicity number (coordination) N, its interatomic distance R and its mean square relative displacement σ^2. In the Equation (1) the S_0^2 term is an empirical parameter taking into account for many body losses while the A(k, R), δ(k, R) and λ are respectively the photoelectron scattering amplitude, phase and mean free path functions [1-3]. These terms can be calculated ab initio, using the scattering theory [1-3] for atomic clusters representative of the local atomic environment around the absorber. It has been demonstrated that the Equation (1) allows describing also the multiple scattering effects, in that case the A(k, R), δ(k ,R) and λ take into account for all the scattering processes along the whole photoelectron path [1, 2].

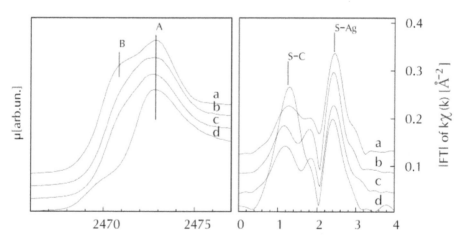

FIGURE 3 S-K edge XAFS on thiolate Ag NPs. Left panel shows the edge region of S-K edge XAFS data collected on nanosized (2–3 nm) Ag particles. The shoulder be signal the presence of Ag₂S phase in the sample. Samples a–c have core shell structure with a large fraction of S sink into the NP forming Ag₂S phase at the NP surface. Sample d has the sharper interface and most of S bridges the Ag ions to thiols chains (S-C bonds). The left panel shows the S-C and Ag-S pair correlation peaks: the higher S-C shell demonstrate the large fraction of S at thiols-NP interface [33].

The features observed in the XANES region are generally stronger than in the EAXFS one because the valence state and coordination chemistry around the absorber produce very specific fingerprints near the Fermi level, which are progressively blurred and confused away from the edge, due to the attenuation of the structural signals and disorder [1-4, 6]. Unfortunately the number of parameters involved, the approximation and the large computation time required make the modeling of XANES data a challenging task, reserved to specialist [1, 2]. Progresses in theory and computation are providing programs for XANES data refinement [25] which are specially useful investigating disordered structures (liquids, amorphous) in which the structural models requires relatively small atomic clusters around the absorber. Efforts are in progress to extend these calculations to long range structures requiring atomic clusters as larger as some 10^2 atoms [26]. Standing these difficulties accurate information can be obtained by comparison with XANES of opportunely selected reference compounds via the linear combination analysis (LCA) which represents a simple and easy way for chemical speciation in complex composite materials [27] as often occur dealing with natural samples in environmental and earth sciences [28], in archaeometry and cultural heritage studies.

2.3 EXAMPLES OF APPLICATIONS TO NANOSTRUCTURED AND COMPOSITE MATERIALS

A number of properties of nanostructured materials derives from their small sizes and/or reduced dimensionality that becoming smaller than the coherence length of the electrons modify the electronic properties and the physico-chemical response of the materials to external solicitations. This gives rise to exotic and unexpected properties charming for fundamental studies and stimulating innovative applications. Understanding the response of nanostructured systems, such as nanoparticles (NPs), thin layers, quantum structures, and so on, is challenging due to several competing factors which affect the strength of the interactions involved: size, dimension, and morphology of the nanostructures, local atomic structure and composition, interactions with matrices, reduced dimensionality, and so on. On the other hand, reducing the particle size progressively reduces the sensitivity of structural investigation techniques, like x-ray diffraction. Therefore, the elemental selectivity and local sensitivity of XAFS are very suited features. As a matter of fact NPs often represent the minority phases in the samples, embedded into supporting matrices: here XAFS selectively probes the NP structure and NPs to matrix interfaces. High quality XAFS data and accurate analysis may provide detailed information on sizes and morphology of the particles [29], may describe the interface structure and reveal the presence of core-shell structures.

Metal NPs and nanostructures may develop a number of peculiar properties relevant for innovative applications. As an example the discover of ferromagnetic (FM) response of thiol capped NPs (size 1–2 nm) of diamagnetic metals like Au, Ag, and Cu, triggered a flurry of activities [30, 31] and a very special surprise was that these NPs may exhibit permanent magnetism even at room temperature while, at such small sizes, Fe, or Co NPs, FM in bulk, became superparamagnetic at low temperatures [27]. This behavior implies huge anisotropy constant (higher that 0.4 eV/atom) which are enormous even if compared with harder magnetic materials. The precise chemical and structural keys leading to ferromagnetism in thiolate NPs of diamagnetic metals are still unclear but two main ingredients are recognized relevant: the particle size, which causes the redistribution of energy levels due to quantum confinement effects, and the charge transfer between metal atoms and the S atoms of thiols chains. The XMCD definitively established the FM behavior of Au and Ag ions in thiolate NPs [31]. On the other hand it has been highlighted as nominally similar samples may depict FM or maybe definitively diamagnetic [32]. In this field XAFS data analysis can supplies very relevant information about the structure at the atomic scale. In particular Ag and Sulfur K edges have provided detailed insight on Ag thiolate NPs [31-elettra report] demonstrating a core-shell structure: a fraction of S remains at the NP surface bridging the thiopls to the NP, a fraction of S sinks into the NP surface giving rise to a Ag-S shell around the Ag fcc core. The Ag-S shell thickness and its composition as function of the synthesis route is likely related to the magnetic response of NPs.

Epitaxial thin films and QD of fcc binary alloys M_xPt_{1-x} (M = Cr, Mn, Fe, and Co) may develop peculiar magnetic properties such as strong PMA making them good candidates for ultra-high density magnetic recording media. The magnetic response of these systems can be varied acting on deposition parameters like: composition, thickness, deposition temperature, and/or rate, nature of substrate and annealing. Polarized XAFS data [15, 16] demonstrated that the anisotropy of chemical order (i.e. the relative arrangement of M and Pt atoms on the fcc lattice) at the very short-range scale (atomic) definitively affects the magnetic properties of these materials. For example in epitaxial $CoPt_3$ thin films [15] and QDs [16], the chemical anisotropy providing Co-Co (Co-Pt) correlations parallel (perpendicular) to the film plane, is strictly correlated with the appearance of PMA.

The XAFS technique can be made depth-selective by generating an x-ray standing wave (XSW) field inside an x-ray waveguides (XWG) [22, 23]. The modulation of XSW field can be modified as a function or reflection angle: tuning the position of antinodes (Figure 4) inside the XWG permits to selectively probe localized structures such as thin layers, films, and buried interfaces, providing details about elemental depth profile and local atomic structure with nanometer depth resolution [34]. This technique has been used to investigate Fe layers embedded in Si and the effect of swift ions irradiation on the evolution of FeSi phases across the interface (Figure 4): in an XWG, made by Si thick layer within two gold surfaces, is placed a thin Fe layer. The Fe fluorescence yield depicts evident maxima (minima) when the antinodes (nodes) of the XSW correspond to the position of Fe layer. The Fe-XAFS spectra were collected at fixed $q = 4\pi \sin(\theta) \lambda$: the data collected at q_B (maximum of the fluorescence yield) mainly describe the Fe structure inside the Fe layer (Fe rich region), data collected at q_A (minimum of the fluorescence yield) mainly describe the Fe structure at the Fe-Si interface and show the occurrence of Si-rich structures.

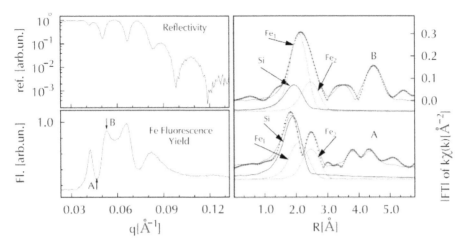

FIGURE 4 Depth selective XAFS spectroscopy. Left panels reflectivity (top) and Fe fluorescence yield (bottom) as a function of the exchanged momentum vector $q=4\pi \sin(\theta)/\lambda$. Maxima (minima) of fluorescence yield signal the antinode (node) corresponding to the position of the Fe layer inside the waveguide.

The XAFS data collected at q_A mainly probe the Fe structure at the Fe-Si interface (Si-rich), data collected at q_B mainly probe the Fe structure inside at Fe layer (Fe-rich).

KEYWORDS

- **Ferromagnetic**
- **Linear combination analysis**
- **Perpendicular magnetic anisotropy**
- **X-ray absorption fine structure**
- **X-ray waveguides**

REFERENCES

1. Rehr, J. J. and Albers, R. C. *Rev. Mod. Phys.*, **72**, 621–654 (2000).
2. D. C. Koningsberger and R. Prins (Eds.). *X-ray absorption: principles*. Wiley, New York (1988).
3. Filipponi, A., Di Cicco, A., and Natoli, C. R. *Phys. Rev. B*, **52**, 15122–15134 (1995).
4. Filipponi, A. and Di Cicco, A. *Phys. Rev. B*, **52**, 15135 (1995).
5. Kothari, D., Reddy, V. R., Gupta, A., Meneghini, C., Aquilanti, G., and Gupta, S. M. *J. Phys. Cond. Mat.*, **22**, 356001–356010 (2010).
6. Mobilio, S. and Meneghini, C. *J. of Non-crystalline solids*, **232**, 25–37 (1998).
7. Cuif, J. P., Dauphin, Y., Doucet, J., Salome, M., and Susini, J. *Geochimica et Cosmochimica Acta*, **67**, 75–83 (2003).
8. Joris, D. J., Janssens, K., Van Der Snickt, G., Van Der Loeff, L., Rickers, K., and Cotte, M. *Anal. Chem.*, **80**, 6436–6442 (2008).
9. Filipponi, A., Di Cicco, A., and De Panfilis, S. *Phys. Rev. Lett.*, **83**, 560–563 (1999).
10. Torchio, R., Pascarelli, S., Mathon, O., Marini, C., Anzellini, S., Centomo. P., Meneghini, C., Mobilio, S., Morley, N. A., and Gibbs, M. R. *J. High Pressure Res.*, **31**, 148–152 (2011).
11. Prestipino, C., Mathon, O., Hino, R., Beteva, A., and Pascarelli, S. *J. Synchrotron Rad.*, **18**, 176–182 (2011).

12. Gawelda, W., Pham, V. T., Benfatto, M., Zaushitsyn, Y., Kaiser, M., Grolimund, D., Johnson, S. L., Abela, R., Hauser, A., Bressler, C., and Chergui, M. *Phys. Rev. Lett.*, **98**, 057401(4) (2007).
13. Zaharko, O. C., Meneghini, C., Cervellino, A., and Fisher, E. *Eur. Phys. J. B*, **19**, 207–213 (2001).
14. Meneghini, C., Di Matteo, S., Monesi, C., Neisius, T., Paolasini, L., and Mobilio, S. *Phys. Rev. B*, **72**, 033111(4) (2005).
15. Meneghini, C., Maret, M., Cadeville, M. C., and Hazemann, J. L. *Journal de Phys. IV*, **7**, 1115–1117 (1997).
16. Liscio, F., Maret, M., Meneghini, C., Mobilio, S., Proux, O., Makarov, D., and Albrecht, M. *Phys. Rev. B*, **81**, 125417(1–9) (2010).
17. Gambardella, P., Rusponi, S.,Veronese, M., Dhesi, S. S., Grazioli, C., Dallmeyer, A., Cabria, I., Zeller, R., Dederichs, P. H., Kern, K., Carbone, C., and Brune, H. Giant. *Science*, **300**, 1130–1133 (2003),
18. Pascarelli, S., Mathon, O., Muñoz, M., Mairs, T., and Susini, J. *J. Synchrotron Rad.*, **13**, 351–358 (2006).
19. Pettifer, R. F., Mathon, O., Pascarelli, S., Cooke, M. D., and Gibbs, M. R. J. *Nature*, **435**, 78–81 (2005),
20. Dalba, G., Daldosso, N., Diop, D., Fornasini, P., Grisenti, R., and Rocca, F. *Journal of Luminescence*, **80**, 103–107 (1998).
21. Stragier, H., Cross, J. O., Rehr, J. J., Sorensen, L. B., Bouldin, C. E., and Woicik, J. C. *Phys. Rev. Lett.*, **69**, 3064–3067 (1992).
22. Gupta, A., Meneghini, C., Saraiya, A., Principi, G., and Avasthi, D. K. *Nucl. Inst. and Met. B*, **212**, 458–464 (2003).
23. Gupta, A., Rajput, P., and Meneghini, C. *Phys. Rev. B*, **75**, 064424(1–6) (2007).
24. López-Flores, V., Ansell, S., Bowron, D. T., Díaz-Moreno, S., Ramos, S., and Muñoz-Páez, A. *Rev. Sci. Instrum.*, **78**, 013109–013120 (2007).
25. Benfatto, M., Congiu-Castellano, G., Daniele, A., and Della Longa, S. *J. Synchrotr. Rad.*, **8**, 267–269 (2001).
26. Monesi, C., Meneghini, C., Bardelli, F., Benfatto, M., Mobilio, S., Manju, U., and, Sarma, D. D. *Phys. Rev. B*, **72**, 174104(1–12) (2005).
27. Torchio, R., Meneghini, C., Mobilio, S., Capellini, G., Garcia-Prieto, A., Alonso, J., Fdez-Gubieda, M. L., Turco Liveri, V., Longo, A., Ruggirello, A. M., and Neisius, T. *J. Magn. and Magn. Mat.*, **322**, 3565–3571 (2010).
28. Lattanzi, P., Meneghini, C., De Giudici, G., and Podda, F. *J. Hazard. Mater.*, **177**, 1138–1144 (2010).
29. Cuenya, B. R., Frenkel, A. I., Mostafa, S., Behafarid, F., Croy, J. R., Ono, L. K., and Wang Q. *Phys. Rev. B*, **82**, 155450(1–8) (2010).
30. P. Zhang, P., and Sham, T. K. *Phys. Rev. Lett.*, **90**, 245502 (2003).
31. Garitaonandia, J. S., Insausti, M., Goikolea, E., Suzuki, M., Cashion, J. D., Kawamura, N., Ohsawa, H., de Muro, I. G., Suzuki, K., Plazaola, F., and Rojo, T. *Nano Lett.* **8**, 661–667 (2008).
32. Goikolea, E., Garitaonandia, J. S., Insausti, M., Gil de Muro, I., Suzuki, M., Uruga, T., Tanida, H., Suzuki, K., Ortega, D., Plazaola, F., and Rojo T. *J. ao Appl. Phys.*, **107**(9), 09E317 (1–3) (2010).
33. Mari, A., Imperatori, P., Marchegiani, G., Pilloni, L., Mezzi, A., Kaciulis, S., Cannas, C., Meneghini, C., Mobilio, S., and Suber, L. *Langmuir*, **26**, 15561–15566, and Elettra experimental report Number: 20100133 (2010).
34. Gupta, A., Darowski, N., Zizak, I., Meneghini, C., Schumacher, G., and Erko, A. *Spectrochimica Acta, part B-Atomic spectroscopy*. **62**, 622–625 (2007).

3 Effect of Ag on Textured Electrical Ceramics

A. Sotelo, M. A. Torres, Sh. Rasekh, L. C. Estepa,
M. A. Madre, and J. C. Diez

CONTENTS

3.1 INTRODUCTION

From the discovery of the first ceramic superconductor [1] and thermoelectric oxides with attractive performances [2], much work has been performed on these materials, leading to the discovery of new families, such as the superconducting $Bi_2Sr_2CaCu_2O_x$ (Bi-2212) [3] or the thermoelectric $Bi_2Sr_2Co_{1.8}O_y$ (BSCO) [4]. These materials are mainly characterized by their good electrical properties in the conducting planes: Cu-O planes in the superconducting and Co-O planes in the thermoelectric ceramics, which are coincident with the crystallographic ab-planes in both cases. This feature can be exploited in the development of commercial applications for these ceramics, with improved electrical properties, when an adequate grain orientation is produced using texturing techniques. The Bi-2212 superconductor and BSCO thermoelectric materials have demonstrated that they are suitable for many practical applications when they are properly processed in order to produce well oriented grains [5, 6]. Among many techniques used successfully to produce well textured materials [7-11], the directional growth from the melt by the laser floating zone (LFZ) method has demonstrated to be a very useful technique for producing well textured Bi-2212 and BSCO rods at high growth rates [12-14]. The materials textured by this technique, have very well aligned crystals, with their c-axis perpendicular to the current flow direction, with low angle tilt grain boundaries, obtaining a good electrical connectivity between grains. Moreover, they show very interesting properties, quantified by the critical current density (J_c) in the case of Bi-2212 and the thermoelectric power factor (PF) for the BSCO materials that allow developing practical applications.

One of the main advantages of the LFZ method is that the materials can be rapidly grown due to the large thermal gradient at the solid-liquid interface [15]. A second additional advantage is the absence of crucible, avoiding external contamination.

However, the poor mechanical characteristics of this kind of materials [16], due to their ceramic nature, impose limitations for practical applications. Some attempts to improve the mechanical properties of Bi-2212 have been performed by Ag addition [17] while no studies have been published, to our knowledge, for the mechanical improvement of BSCO.

The aim of this work is to study the effect of Ag addition on the Bi-2212 and BSCO systems processed by LFZ. The changes produced on the microstructure have been related with the measured electrical and mechanical properties of both families.

3.2 EXPERIMENTAL

The Bi-2212/x wt% Ag and BSCO/x wt% Ag (x = 1, 3, 5, and 10) precursors have been prepared from commercial Bi_2O_3 (Panreac, 98 + %), $SrCO_3$ (Panreac, 98 + %), $CaCO_3$ (Panreac, 98.5 + %), CuO (Panreac, 98 + %), $Co(NO_3)_2 \cdot 6 H_2O$ (Aldrich, 99%) and Ag (Aldrich, 99.9 + %) powders by a sol-gel method *via* nitrates [18] to assure total cation solution, small particle size, and good homogeneity in the initial mixture. The initial powders were suspended in distilled water, followed by dropwise addition of HNO_3 (Panreac, analysis grade) until a clear blue (for Bi-2212) or pink (for BSCO) solution was obtained. Citric acid (Panreac, 99. 5%) and ethylene glycol (Panreac, 99%) were added to this solution in the adequate proportions. Evaporation of the solvent was performed slowly in order to decompose the nitric acid excess, which allows the polymerization reaction between ethylene glycol and citric acid, forming a light blue (for the Bi-2212) or light pink (for the BSCO) gel. The dried product was then decomposed (producing a slow self-combustion) by heating at 650K for 1 hr. The decomposed solid was mechanically ground and calcined at 1025 and 1075K for 12 hr, with an intermediate grinding. This calcination process has been fixed [19] to specifically decompose the alcaline earth carbonates avoiding their decomposition in the LFZ process, which would lead to bubble formation inside the melt, disturbing the crystallization front. From these powders, green cylindrical ceramic precursors were prepared by isostatic pressing at about 200 MPa for 1 min.

The obtained cylinders were then used as feed in a directional solidification process performed in an LFZ installation [20]. The textured bars were obtained using a continuous power Nd: YAG laser (λ = 1,064 nm), under air, at a growth rate of 30 mm/

hr and a relative rotation of 18 rpm between seed and feed. Using these growth conditions and adjusting the laser power input to obtain a molten zone of 1–1.5 times the rod diameter, it is possible to obtain a stable solidification front, which allows the fabrication of homogeneous textured bars. After the texturing process, some differences are arising between the two types of ceramics:

(1) The Bi-2212 ceramics present incongruent melting and, in consequence, after the directional solidification process, it is necessary to perform a thermal treatment in order to form the Bi-2212 superconducting phase [21, 22]. This annealing process was performed under air, and consisted in two steps: 60 hr at 1125K, followed by 12 hr at 1075K and, finally, quench in air to room temperature. Before the thermal treatment, silver contacts were painted on the as-grown samples for the electrical measurements. After annealing, the silver contacts have typical resistance values 1 μΩ.

(2) After the directional growth process, thermoelectric ceramics are composed mainly by the thermoelectric phase, together with some secondary phases, with good thermoelectric properties [14, 23]. As a consequence, no further treatments are necessary for this type of materials.

Structural identification of all ceramic samples was performed by powder XRD utilizing a Rigaku D/max-B X-ray powder diffractometer (CuKα radiation) with 2θ ranging between 10 and 60 degrees. Microstructural characterization was performed on polished longitudinal cross-sections of the samples, in a scanning electron microscope (SEM, JEOL JSM 6,400) equipped with an energy dispersive spectroscopy (EDX) system.

Mechanical characterization has been performed by flexural strength, using the three-point bending test in an Instron 5,565 machine with a 10 mm loading span fixture and a punch displacement speed of 30 μm/min.

Electrical properties have been determined in different ways for both types of ceramics:

(1) For the Bi-2212/Ag composite materials, annealed samples were, approximately, 4 cm long and were measured using the standard four-probe configuration. Critical current density (J_c) values were determined at 77K using the 1 μV/cm criterion. Resistivity as a function of temperature, from 77 to 300K, was measured using a dc current of 1 mA, in order to determine the transport T_c values.

(2) In the case of the BSCO/Ag composites, electrical resistivity (ρ) and thermo-electric power (TEP) were simultaneously determined by the standard dc four-probe technique in a LSR-3 measurement system (Linseis GmbH). They were measured in the steady state mode at temperatures ranging from approximately 300 to 950K under He atmosphere. With the electrical resistivity and TEP data, the power factor (PF = $[TEP]^2/\rho$) has been calculated in order to determine the samples performance.

3.3   DISCUSSION AND RESULTS

3.3.1 As-grown Materials Microstructural Characterization

The SEM microstructural observations on as-grown materials showed that all the textured rods are multiphasic for each group of samples (Bi-2212/Ag and BSCO/Ag), indicating that in both cases they have an incongruent melting. Moreover, the secondary phases identified in each family are the same, independently of Ag content, indicating that the melt is chemically similar in each family and that Ag produces minor effects on the solidification process. A representative micrograph of each type of material, performed on longitudinal polished sections of the as-grown materials is displayed in Figure 1. For the Bi-2212/3 wt% Ag ceramic (Figure 1(a)), five different phases can be observed and identified by EDX. The primary solidification phase is the Bi-free $(Ca, Sr)CuO_2$ (dark contrast in the Figure 1(a), and marked as #1). This phase appears well aligned with the rod axis, as expected from the flat solidification interface observed in the growth process [24]. Light gray phase (marked as #2 in Figure 1(a)) shows a composition close to the ideal $Bi_2Sr_2CuO_x$ (Bi-2201) stoichiometry. This phase's formation is promoted by its fast solidification kinetics from the Bi-enriched liquid, a process that follows initial nucleation of $(Ca, Sr)CuO_2$. Moreover, from the trends observed in Figure 1(a) it can be deduced that the grain growth of this Bi-rich phase mainly follows the alignment of the primary solidified phase. Gray phase (marked as #3 in Figure 1(a)) has a composition close to the stoichiometric Bi-2212 with a slightly lower Ca-content, corresponding to Bi-2212/Bi-2201 intergrowths, usually appearing between Bi-2201 grains. Small black spots (marked as #4 in Figure 1(a)) distributed inside the Bi-rich phase have been identified as CaO particles. Finally, #5 in Figure 1(a) is indicating Ag particles, which can be also seen as gray contrasts but can be distinguished by their shapes, as they tend to be spherical.

Similar microstructure is found for the BSCO/Ag ceramics, which is shown in Figure 1(b) for the BSCO/5 wt% Ag samples. In these samples, five main phases, as well as Ag particles (numbered as #5 in Figure 1(b)) are found. In these samples, Ag tends to change its shape from the spherical one found in Bi-2212/Ag rods to a more elliptical one trying to fill the free space between the alternating thermoelectric grains layers. The major phases found are the gray phase (#6 in Figure 1(b)), corresponding to $Bi_2Sr_2Co_2O_y$, and the light gray contrast (#7 in Figure 1(b)), identified as $Bi_2Sr_2CoO_y$, which appear as well aligned alternated layers. Minor phases appear as white contrast (#8 in Figure 1(b)) for the Co-free one, black one (#9 in Figure 1(b)) for CoO, and dark gray (#10 in Figure 1(b)), with a composition close to the thermoelectric $Sr_2Co_{2.8-x}Bi_xO_y$ one [25].

Another important feature arising from SEM observations is that for all samples with 5 wt% Ag, big Ag particles are found, as illustrated by Figure 2 where a representative transversal polished section of BSCO/5 wt% Ag is displayed. From these observations, it is clear that the Ag solubility is very low in both families.

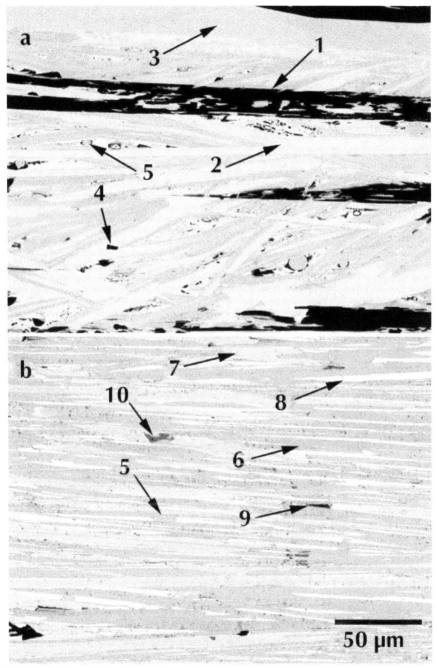

FIGURE 1 The SEM micrographs performed on longitudinal polished sections of the as-grown materials: (a) Bi-2212/3 wt% Ag and (b) BSCO/5 wt% Ag. The different phases are: (1) $(Ca,Sr)CuO_2$, (2) Bi-2201, (3) Bi-2212/Bi-2201 intergrowths, (4) CaO, (5) Ag, (6) $Bi_2Sr_2Co_2O_y$, (7) $Bi_2Sr_2CoO_y$, (8) Co-free, (9) CoO, and (10) $Sr_2Co_{2.8-x}Bi_xO_y$.

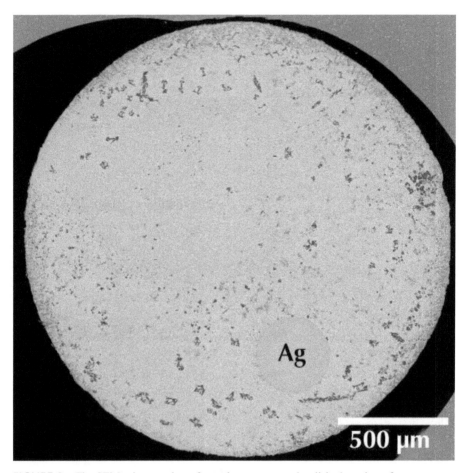

FIGURE 2 The SEM micrograph performed on transversal polished section of an as-grown BSCO/5 wt% Ag sample showing a big Ag particle formed in the solidification process.

From the observations, it is clear that Bi-2212/Ag samples must be annealed in order to promote the formation of the superconducting Bi-2212 phase while it is not necessary in the case of the BSCO/Ag composites as in previous works it has been demonstrated that these as-grown materials posses interesting thermoelectric properties [14]. As a consequence, Bi-2212/Ag samples will be studied after annealing while BSCO samples will be characterized in the as-grown state.

3.3.2 XRD Characterization

The powder XRD patterns for the different BSCO/Ag composite samples, grown by the LFZ process, are displayed in Figure 3. From these data, it is clear that all the samples have very similar diffraction patterns and show minor peaks corresponding to non-thermoelectric secondary phases. The highest peaks belong to the misfit cobaltite phase and are in agreement with reported data [26]. Weak diffraction peaks (marked with a ▼) are related with the solid solution Bi-Sr-O secondary phases, the ♦ indicates the Si (111) diffraction peak, used as reference, and # marks the (111) peak of Ag, which is difficult to observe as it is appearing as a shoulder on the ~38.5° diffraction peak.

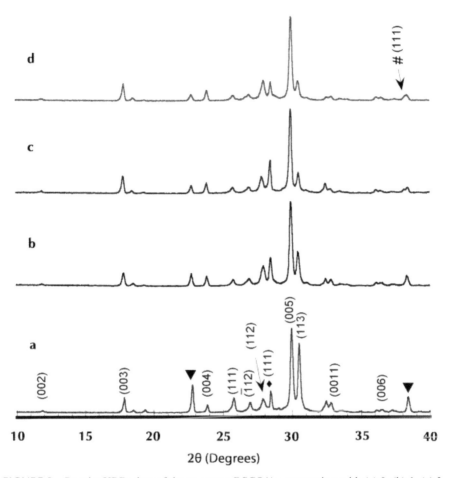

FIGURE 3 Powder XRD plots of the as-grown BSCO/Ag composites with (a) 0, (b) 1, (c) 3, and (d) 5 wt% Ag. ▼ Bi-Sr-O secondary phases, ♦ Si used as reference, and # Ag.

For the Bi-2212/Ag materials, the powder XRD plots have been recorded after the annealing process performed at 1,125K for 60 hr to produce the Bi-2212 phase formation. As it can be clearly seen in Figure 4, the diffraction patterns are very similar for all the samples, independently of the Ag content, showing that they are composed by nearly pure Bi-2212 phase. Major peaks are associated to the Bi-2212 phase, accompanied by small amounts of Bi-2201 (indicated by * in Figure 4). On the other hand, Ag (marked with #) is only detected for samples with 3 or 5 wt% Ag.

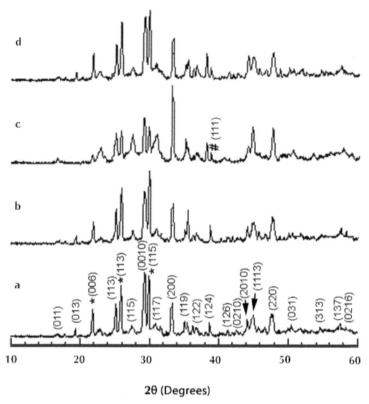

FIGURE 4 Powder XRD plots of the annealed Bi-2212/Ag composites with (a) 0, (b) 1, (c) 3, and (d) 5 wt% Ag. * Bi-2201 secondary phases, and # Ag.

3.3.3 Microstructural Characterization of Annealed Bi-2212/Ag Composites

When observing SEM micrograph presented in Figure 5 (corresponding to a Bi-2212/3 wt% Ag transversal polished section), it is clear that major phase is Bi-2212 (gray contrast) as it was found on the XRD analysis. In contrast, there are still some minor secondary phases identified by EDX as Bi-2201 (white contrast), and (Sr, Ca)CuO$_2$ (dark gray contrast). Moreover, Ag particles can also be found, with a different shape than they showed in the as-grown samples. This shape change of the Ag particles is due to the formation of a liquid phase [27] and its migration to the elongated holes between superconducting grains, originated in the Bi-2212 formation process from Bi-2201 and other secondary phases. The number and size of these silver inclusions increases with Ag content leading, in the case of the 5 wt% Ag samples, to the formation of big spherical particles (already observed in Figure 2). Due to the high stability of these big spherical particles, they do not change their shape in the relatively short annealing time, as it is illustrated with Figure 6 where an annealed Bi-2212/5 wt% Ag transversal polished section is presented. In this figure it is possible to see a slight plastic deformation from the spherical shape due to the superconducting grains growth during the thermal treatment. Moreover, these Ag particles induce the misalignment of their superconducting surrounding grains.

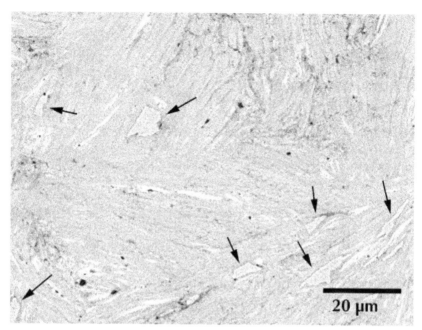

FIGURE 5 The SEM micrograph performed on transversal polished section of the annealed Bi-2212/3 wt% Ag showing the different phases, gray contrast corresponds to Bi-2212, white contrast to Bi-2201, and dark gray to $(Sr,Ca)CuO_2$. Ag particles are indicated by arrows.

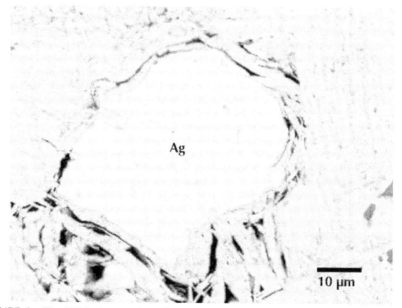

FIGURE 6 The SEM micrograph performed on transversal polished section of the annealed Bi-2212/5 wt% Ag showing a big Ag particle which does not change its shape during the annealing process.

3.3.4   Mechanical Characterization

To study the mechanical behavior, flexural strength tests were made on as-grown BSCO/Ag and annealed Bi-2212/Ag samples, using at least six samples for each composition.

In Figure 7 it is represented the flexural strength values as a function of Ag content for the as-grown BSCO/Ag samples together with their standard error. At first sight it is clear that the Ag addition improves mechanical properties, compared with the pure samples, in an important manner independently of the Ag amount. Moreover, Ag addition reduces the standard error values in all cases. The best mechanical strength values are obtained for small Ag contents due to the small size and good distribution of Ag particles all over the ceramic matrix, producing an increase of about 25% from the values obtained for the pure Bi-2212. This effect is produced by the Ag particles found between superconducting platelets which provide a plastic flow region, reducing crack propagation [17]. The progressive decrease of flexural strength for higher Ag contents can be associated to the Ag particles size increase, reducing their ability to adapt their shape to that of the holes between the superconducting grains.

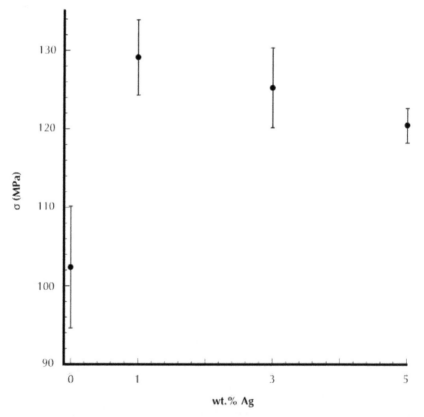

FIGURE 7   Flexural strength for the as-grown BSCO/Ag samples as a function of the Ag content. The error bars indicate the obtained standard error.

For the Bi-2212/Ag annealed samples, a very similar behavior is found. As it can be seen in Figure 8, the best results are obtained for small Ag additions, where an increase of about 40% has been obtained for samples containing 1 wt% Ag. Moreover, as it was observed and explained for higher silver contents, flexural strength decreases significantly.

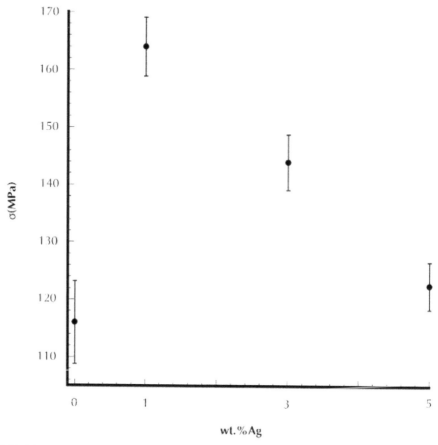

FIGURE 8 Flexural strength for the annealed Bi-2212/Ag samples as a function of the Ag content. The error bars indicate the obtained standard error.

3.3.5 Thermoelectric Characterization of As-grown BSCO/Ag Composites

The temperature dependence of the resistivity, as a function of Ag content is shown in Figure 9. The $\rho(T)$ curves show an important decrease of the resistivity with Ag addition, which must be related to the improved TE grains electrical connectivity, as this behavior cannot be explained using a resistivity mixture rule [28]. The minimum value at room temperature is around 200 $\mu\Omega$m for high Ag content (3 and 5 wt% Ag), about the best reported values for this type of textured materials measured on the ab plane at around 550K [29].

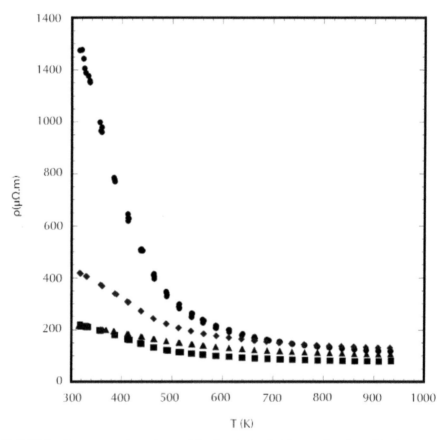

FIGURE 9 Temperature dependence of the electrical resistivity as a function of the Ag content in as-grown BSCO/x wt% Ag samples, for x = 0 (●), 1 (♦), 3 (■), and 5 (▲).

Figure 10 shows the variation of the TEP with the temperature as a function of Ag content. The sign of the TEP is positive for the entire measured temperature range, which confirms a mechanism involving a predominant whole conduction mechanism. The values of the TEP show two different behaviors, one for samples with Ag contents ≤1 wt% and the other for higher Ag contents. In the first case, TEP is decreasing from room temperature to around 775K and then increases while for high Ag additions, TEP is increasing from room temperature to around 575K and decreasing for higher temperatures. The maximum value at room temperature is reached for the 5 wt% Ag samples (about 150 µV/K) which is higher than those reported for solid state sintered materials (about 120 µV/K) or textured ones (about 140 µV/K). The high TEP value obtained for these samples can be explained by the high number of oxygen vacancies generated on the LFZ growth process which are probably produced in larger amounts than in samples synthesized in a classic solid state reaction. As a consequence, in order to maintain the electrical neutrality, a reduction of Co^{4+} to Co^{3+} is produced, leading to the decrease of holes concentration [30]. It has already been evidenced that, under reduced conditions, the misfit phase $[Ca_2CoO_3]$ $[CoO_2]_{1.62}$ (close to the present misfit system) contains considerable amounts of oxygen vacancies [31].

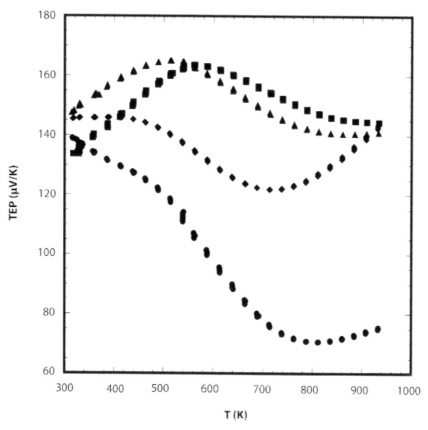

FIGURE 10   Temperature dependence of the thermopower as a function of the Ag content in as-grown BSCO/x wt% Ag samples, for x = 0 (●), 1 (♦), 3 (■), and 5 (▲).

In order to evaluate the thermoelectric performance of these materials, the PF has been calculated. The temperature and Ag dependences of PF are calculated from the resistivity and TEP data and displayed in Figure 11. The PF at room temperature is increased, from 0.02 to 0.10 mW/K²m, when the Ag contents are increased from 0 to 5 wt%. This maximum value is higher by one order of magnitude when compared to those measured in randomly oriented grains ceramics (equal to 0.01 mW/K²m), and comparable to those reported at around 550K by Itahara et al. using the reactive templated grain growth (RTGG) technique and measuring on the ab plane [29].

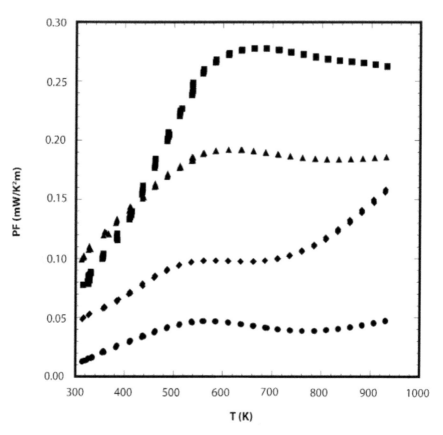

FIGURE 11   Temperature dependence of the power factor as a function of the Ag content in as-grown BSCO/x wt% Ag samples, for x = 0 (●), 1 (◆), 3 (■), and 5 (▲).

3.3.6   Electric Characterization of Annealed Bi-2212/Ag Composites

Figure 12 shows the temperature dependence of the resistivity as a function of Ag content for annealed Bi-2212/Ag composites. In this figure, it can be seen that T_c is the same for all the samples indicating the same intragranular behavior in all cases. As a consequence it can be deduced that Ag is not incorporated in the superconductor crystalline structure. Moreover, the high temperature resistivity values are very similar for all the samples indicating good electrical grains connectivity even when Ag is not present. The slight increase on the high temperature resistivity found for the 5 wt% Ag samples can be probably associated to the superconducting grain misalignment around the big Ag particles, as observed in Figure 6.

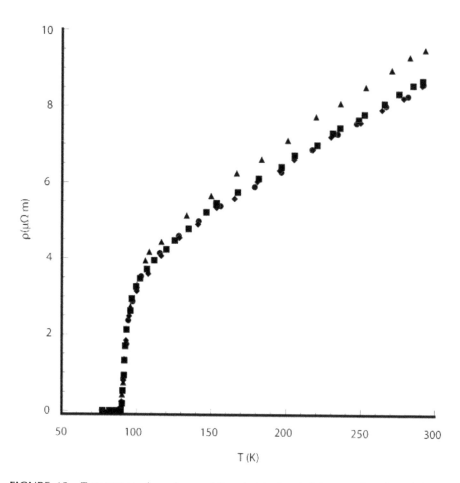

FIGURE 12 Temperature dependence of the electrical resistivity as a function of the Ag content in annealed Bi-2212/x wt% Ag samples, for x = 0 (●), 1 (♦), 3 (■), and 5 (▲).

The evolution of the J_c values as a function of the Ag content at 77K is shown in Figure 13. As it can be clearly seen in this figure, J_c values higher than 3,500 A/cm² are obtained for Ag added samples. The low value found for the undoped sample is due to the thermal treatment which is not the optimal one for pure Bi-2212 compositions. When the undoped samples are properly treated, they reach, approximately, the same critical current values [32]. These results confirm that Ag addition has not a negative effect on the electrical transport properties. In fact, for the prepared samples (about 2 mm diameter) is possible to get electrical intensities as high as around 100 A without resistive loses.

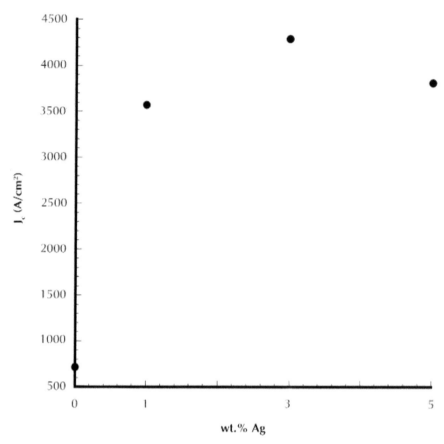

FIGURE 13   Critical current density values at 77K as a function of the Ag content in annealed Bi-2212/x wt% Ag samples.

3.4   CONCLUSION

It has been demonstrated that it is possible to obtain textured Bi-2212/Ag and BSCO/Ag advanced ceramic composites, through a directional growth process from the melt by the LFZ melting technique. This technique can be useful to obtain improved properties on materials with high anisotropic properties, as Bi-based ceramic superconductors and Co-based thermoelectric oxides. In all cases, Ag addition improves flexural strength by providing a plastic-flow region which reduces crack propagation. It has been found that for this layered systems, the optimal Ag addition is about 1 wt% Ag appears as small spherical or ellipsoidal particles between the plate like grains for the as-grown or annealed samples, respectively. Increasing Ag content leads to bigger precipitates which can produce a detrimental effect on their mechanical properties.

From the electrical point of view, it has been also shown that Ag addition improves grains connectivity in the case of the BSCO/Ag composites. In these materials a very important reduction on the resistivity values at room temperature has been obtained, from 1,300 to 200 μΩm with additions of 3 wt% Ag. In the case of Bi-2212/Ag composites, Ag addition does not affect significantly superconducting grains connectivity.

KEYWORDS

- **Laser floating zone method**
- **Textured electrical ceramics**
- **Texturing techniques**
- **Thermoelectric ceramics**
- **Thermoelectric power**

ACKNOWLEDGMENTS

The authors wish to thank the Spanish Ministry for Science and Innovation (MICINN-FEDER, MAT2008-00429 and MAT2008-05983-C03-01 projects) and the Gobierno de Aragón (Research Groups T12 and T72) for financial support. The technical contributions of J. A. Gómez and C. Gallego are also acknowledged.

REFERENCES

1. Bednorz, J. G. and Müller, K. A. Z. *Phys. B-Condensed Matter*, **64**, 189–193 (1986).
2. Terasaki, I., Sasago, Y., and Uchinokura, K. *Phys. Rev. B*, **56**, 12685–12687 (1997).
3. Maeda, H., Tanaka, Y., Fukutomi, M., and Asano, T. *Jpn. J. Appl. Phys. Lett.*, **27**, 209–210 (1988).
4. Unahashi, R., Matsubara, I., and Sodeoka, S. *Appl. Phys. Lett.*, **76**, 2385–2387 (2000).
5. Herrmann, P. F. *Current leads. In Handbook of Applied Superconductivity*. B Seeber (Ed.). IOP Publishing Ltd., Bristol and Philadelphia, pp. 801–843 (1998).
6. Noe, M., Juengst, K. P., Werfel, F. N., Elschner, S., Bock, J., Breuer, F., and Kreutz, R. *IEEE Trans. Appl. Supercond.*, **13**, 1976–1979 (2003).
7. Garnier, V., Caillard, R., Sotelo, A., and Desgardin, G. *Physica C*, **319**, 197–208 (1999).
8. Martin-Gonzalez, M. S., Garcia-Jaca, J., Moran, E., and Alario-Franco M. A. *J. Mater. Res.*, **14**, 3497–3505 (1999).
9. Maeda, H., Ohya, K., Sato, M., Chen, W. P., Watanabe, K., Motokawa, M., Matsumoto, A., Kumakura, H. and Schwartz, J. *Physica C*, **382**, 33–37 (2002).
10. Zhang, Y. F., Zhang, J. X., and Lu, Q. M. *Ceram. Int.*, **33**, 1305–1308 (2007).
11. Seabaugh, M. M., Kerscht, I. H., and Messing, C. L. *J. Am. Ceram. Soc.*, **80**, 1181–1188 (1997).
12. Feigelson, R. S. Gazit, D., Fork, D. K., and Geballe, T. H. *Science*, **240**, 1642–1645 (1988).
13. Sotelo, A., Madre, M. A., Diez, J. C., Rasekh, Sh., Angurel, L. A., and Martınez, E. *Supercond. Sci. Technol.*, **22**, 034012 (2009).
14. Diez, J. C., Guilmeau, E., Madre, M. A., Marinel, S., Lemmonier, S., and Sotelo, A. *Solid State Ionics*, **180**, 827–830 (2009).
15. Revcolevschi, A. and Jegoudez, J. *Prog. Mater. Sci.*, **42**, 321–339 (1997).
16. Pastor, J. Y., Poza, P., and Llorca, J. *J. Am. Ceram. Soc.*, **82**, 3139–3144 (1999).
17. Joo, J., Singh, J. P., Warzynski, T., Grow, A., and Poeppel, R. B. *Appl. Supercond.*, **2**, 401–410 (1994).
18. Ruiz, M. T., De la Fuente, G. F., Badía, A., Blasco, J., Castro, M., Sotelo, A., Larrea, A., Lera, F., Rillo, C., and Navarro, R. *J. Mater. Res.*, **8**, 1268–1276 (1993).
19. Sotelo, A., De la Fuente, G. F., Lera, F., Beltrán, D., Sapiña, F., Ibáñez, R., Beltrán A., and Bermejo, M. R. *Chem. Mater.*, **5**, 851–856 (1993).

20. Carrasco, M. F., Costa, F. M., Silva, R. F., Gimeno, F., Sotelo, A., Mora, M., Diez, J. C., and Angurel, L. A. *Physica C*, **415**, 163–171 (2004).
21. Funahashi, R., Matsubara, I., Ueno, K., and Ishikawa, H. *Physica C*, **311**, 107–121 (1999).
22. Mora, M., Martinez, E., Diez, J. C., Angurel, L. A., and de la Fuente, G. F. *J. Mater. Res.*, **15**, 614–620 (2000).
23. Sotelo, A., Guilmeau, E., Madre, M. A., Marinel, S., Lemmonier, S., and Diez, J. C. *Bol. Soc. Esp. Ceram. V.*, **47**, 225–228 (2008).
24. De la Fuente, G. F., Diez, J. C., Angurel, L. A., Peña, J. I., Sotelo, A., and Navarro, R. *Adv. Mater.*, **7**, 853–856 (1995).
25. Klein, Y., Pelloquin, D., Hebert, S., Maignan, A., and Hejtmanek, J. *J. Appl. Phys.*, **98**, 013701 (2005).
26. Xu, G. J., Funahashi, R., Shikano, M., Matsubara, I., and Zhou, Y. Q. *J. Appl. Phys.*, **91**, 4344–4347 (2002).
27. Majewski, P., Sotelo, A., Szillat, H., Kaesche, S., and Aldinger, F. *Physica C*, **275**, 47–51 (1997).
28. Kasap, S. O. *Principles of electronic materials and devices*. McGraw-Hill, New York (2006).
29. Itahara, H., Xia, C., Sugiyama, J., and Tani, T. *J. Mater. Chem.*, **14**, 61–66 (2004).
30. Maignan, A., Pelloquin, D., Hebert, S., Klein, Y., and Hervieu, M. *Bol. Soc. Esp. Ceram. V.*, **45**, 122–125 (2006)
31. Karppinen, M., Fjellvag, H., Konno, T., Morita, Y., Motohashi, T., and Yamauchi, H. *Chem. Mater.*, **16**, 2790–2793 (2004).
32. Angurel, L. A., Diez, J. C., Martinez, E., Peña, J. I., De la Fuente, G. F., and Navarro, R. *Physica C*, **302**, 39–50 (1998).

4 Nanoparticles Filled Epoxy-based Adhesive for *in Situ* Timber Bonding

Z. Ahmad, M. P. Ansell, D. Smedley, and P. Md Tahir

CONTENTS

4.1   INTRODUCTION

Rods, dowels, and plates are bonded into timber with high strength adhesives for timber connections and are widely used for repair of timber structures [1-4]. Rods or plates made from pultruded fiber reinforced plastic (FRP) or metal are bonded into timber using structural adhesives. The adhesives are back-injected into oversize holes or slots and the rods or plates are pushed into the adhesives forming a void-free inter-face and the adhesive is allowed to cure. The bond integrity influences the ultimate capacity of the bonded structure as well as serviceability and depends on a number of factors including the type of adhesive and adherend, cure cycle, bondline thickness, and the environment.

Structural adhesives and adhesive bonding are reviewed by Comyn [5], Kinloch [6], and Mays and Hutchinson [7]. Studies on bonded-in connections are still limited and the focus of these studies is mainly on the design of the connections with varia-tion in rod type and size, glueline thickness, type of timber and type of connection [8-11]. The bond strength depends on how well the adhesive wets the adherends and the lower the viscosity, the more easily the adhesive will wet the adherend. Carpenter [12] provides an overview of strength theories for the design of adhesively bonded lap joints with composite adherends and numerous references review adhesive bonding of timber [13, 14]. However, these references focus on low viscosity adhesives which require pressure and high temperatures for curing. In order to achieve effective room temperature *in situ* bonding of pultrusions into timber without the application of pres-sure, the adhesives used must be thixotropic and shear thinning so that they cannot run out of inverted holes. As adhesives are expected to hold the timber materials together, the right choice of adhesive suitable for the job is very important. There are many adhesives available however none of the adhesives has dominated the market place.

Epoxy-based adhesives have special chemical characteristics compared with other thermosetting resins: no byproducts or volatiles are formed during curing reactions, so shrinkage is low, epoxy resins can be cured over a wide range of temperatures, and the degree of cross-linking can be controlled. Depending on the chemical structure of the curing agents and on curing conditions, the properties of cured epoxy resins are versatile, including excellent chemical and heat resistance, high adhesive strength, low shrinkage, good impact resistance, high strength, hardness, and high electrical insulation.

But cured epoxy systems have one main drawback: their considerable brittleness, which results in poor fracture toughness poor resistance to crack propagation, and low impact strength. This inherent brittleness has limited their application in fields requir-ing high impact and fracture strengths, such as reinforced plastics, matrix resins for composites, resins for bonded-in structure and coatings. Therefore, in the last few de-cades, much attention has been given to improving the thermal and mechanical prop-erties of epoxy resins, especially in making them tougher. It is widely believed that the brittleness of epoxy resins is associated with their highly cross-linked structures which absorb insignificant amounts of energy during the fracture process. Several ap-proaches have been used to enhance the toughness of epoxy resins which include:

- Chemical modification of the epoxy backbone to make it more flexible structure
- Increasing the molecular weight of the epoxy
- Lowering the cross-link density of the matrix
- Incorporation of a dispersed toughener phase in the cured polymer matrix
- Incorporation of inorganic fillers into the neat resin

• This investigation focused on the third approach by adding fillers including nanosilica, liquid rubber, and microsized particles such as quartz, bentonite, and mica.

The matrix material plays an important role in filled adhesives systems. If the adhesive matrix has high ductility, the addition of fillers can increase the toughness of composite. Epoxy itself has poor impact properties. The incorporation of liquid rubber into epoxy adhesives has been shown to produce stable, compatible adhesives which develop two-phases during cure [15-17]. Earlier in 1970, McGarry [18] showed that the addition of liquid rubbers had a significant improvement in the fracture toughness of the modified epoxy adhesives. In terms of bonding strength, Ratna and Banthia [19] reported a two-fold increase in lap shear strength using carboxyl-terminated poly (2-ethylhexyl acrylate) (CTPEHA) as the liquid rubber.

The incorporation of nanofillers into thermosets has also attracted considerable interest indicated by the recent increase in the number of publications [20-22]. Incorporation of superfine inorganic particles play a remarkable role in modifying epoxies due to their smaller dimension, larger specific area, higher surface energy, and better surface reaction [23] found that the addition of silica nanoparticles up to 10 wt% brings about a considerable increase in fracture toughness.

Moisture is commonly encountered in the service environment and must be considered a critical factor in determining the long term reliability of adhesively bonded joints. It is well known that warm and moist surroundings can lead to a considerable loss in strength of adhesive joints and this reduction can be due to the ingress of water which may be related to physical and chemical modification occurring at the interface or inter phase between the substrate and adhesive [24] or the degradation of the adhesive and/or the substrate. Moisture absorbed in a polymer matrix can lead to a wide range of effects, both reversible and irreversible, including plasticization by weakening the intermolecular interaction among the functional groups of the chain [25], structural damage such as microcavities or crazes [26, 27], further cross-linking [28] and chemical degradation of the matrix due to hydrolysis and oxidation during long term exposure to water [29]. Cross-linked epoxy adhesives can absorb water up to a maximum of about 10% moisture by mass [30], depending on the chemical composition, stress state, exposure time, and temperature. For an epoxy system, it is estimated that the critical water concentration at which adhesion is significantly affected is about 1.35–1.45%, and the critical relative humidity is around 50–65% [30-32]. Antoon and Koenig [29] investigated the effects of moisture on anhydride-cross-linked epoxy resin films by means of fourier transform infrared (FTIR) spectroscopy and reported that slow oxidation processes occur in air at 100% relative humidity.

Most of the work mentioned was done on low viscosity epoxy adhesives. For glued-in timber connections, high viscosity epoxy-based adhesives are required so that thicker glue line can be produced. Harvey and Ansell [11] studied glued-in glass fiber reinforced plastic (GFRP) rods used in beam to column connections with glue-line thickness ranging from 2–4.5 mm using epoxy-based adhesives including CB10TSS and found that the joint with CB10TSS showed a high ductility with higher glue-line thickness. Therefore, in this study, CB10TSS formulation was modified to incorporate the addition of nanorubber and microparticles (quartz, mica, and bentonite) with the hope to improve the mechanical and thermal properties and environmental stability.

Research on bonded-in timber connections has mainly been conducted to perfect the design of joints. However, less attention has been paid to the properties of the adhesives used for bonding-in connections *in situ* on the construction site. Adhesives used must be thixotropic, environmentally stable and friendly, applied at without pressure. A wide range of tests have been conducted to examine the thermal, physical, and mechanical properties of three thixotropic and room temperature cured epoxy-based adhesives formulated specifically for *in situ* timber bonding, which is a mixture of diglycidylether of bisphenol-A (DGEBA) with reactive diluent glycidlyether (monofunctional), silica fume particles and hardener, a mixture of polyetheramines and other ingredients. Albipox is a modification of CB10TSS with the addition of nanodispersed carboxyl-terminated butadiene acrylonitrile (CTBN) and timberset is an adhesive formulation containing ceramic microparticles (quartz, mica, and bentonite) and cured with an aliphatic diamine (Trimethyl hexamethylene diamine). Reinforcement of CB10TSS with phase separated nanorubber particles CTBN remarkably improves the tensile, flexural and interlaminar shear strength and glass transition temperature of the standard adhesive. However, the ceramic particles addition significantly raised modulus of elasticity (MOE) but embrittled the adhesive. The addition of only 2.0% CTBN raised the Tg of CB10TSS by 11°C. The same trend is also observed for timberset. The addition of microfillers to CB10TSS improves the MOE and glass transition temperature (improves the Tg by 22°C) but reduces the flexural, tensile and compressive strengths. Scanning electron microscopy (SEM) and transmission electron microscopy (TEM) provide evidence of excessive shear yielding and a clear phase separation in the rubber filled adhesive respectively. These well-defined, finely divided nanostructures in Albipox result in higher strength than the timberset with microparticle filled adhesive. The environmental stability of epoxy-based adhesives was evaluated following soaking in water for up to 90 days. Nanofiller additions enhance environmental stability but the addition of microparticles provides better moisture resistance.

4.2 METHODOLOGY

4.2.1 Materials

Three types of adhesives used in this study. The first type was a low viscosity two-phase epoxy resin system consisting of Part-A (epoxy resin mixture of DGEBA with reactive diluents glycidlyether (monofunctional) and silica fume particle) and Part-B (hardener, mixture of polyetheramines) and this mixture denoted as CB10TSS and is considered neat adhesive. The other two adhesives were formulated from neat adhesive but with the addition of (a) liquid rubber of CTBN and (b) microceramics particles. These adhesives are designated as albipox and timberset respectively.

4.2.2 Specimen Preparation

The adhesives were manually mixed at a constant speed so as not to form a vortex as this might have entrapped air. The vigorous mixing was avoided as this can produce highly reactive volatile vapor bubbles which could affect the final product by creating voids. The mixing was continued for about 15 min until the two components were well blended. After mixing the two components the adhesive was placed in a vacuum chamber for about 5 min in an attempt to release air bubbles. After that the mixture was transferred into a plastic rectangular mould of 2 mm thick, 500 mm wide, and 500 mm long coated with release agent. Another glass plate was placed on top of the mould and weights were added to apply some pressure in order to achieve a flat surface. The adhesive was left in the mould for 6 days to cure. After demoulding, the adhesive plate was cut by diamond saw for the bending test specimens in accordance with the British Standard BS EN ISO 178 (Plastic-Determination of flexural properties) (2003). After cutting the specimens from the plates, 50 specimens were randomly selected and divided into five groups with ten specimens per group for bending tests at different times following curing: 7, 10, 20, 30, and 40 days to study the effect of curing time on flexural strength. The remainder of the plate was cut for dynamic mechanical thermal analysis (DMTA) specimens. The tensile specimen was prepared in the aluminum mould for the dumbbell-shaped. The compression specimens were cut from the adhesive panel into a size of 12.5 × 12.5 × 25 mm prisms according to ASTM D695 (1991): compressive properties of rigid plastic.

4.2.3   Test Methods
A series of test were conducted and described as follows:

Flexural Test
Flexural tests under three-point load method were performed according to BS EN ISO 178:2003. The tests were conducted using an Instron 1122 universal test machine at a crosshead speed of 1 mm/min.

Tensile Test
The tensile properties (Young's modulus and tensile strength) of the adhesive were determined using an Instron 1185 universal testing instrument equipped with a 100 kN load cell by securing the tensile samples in the jaws of the instrument. A 25 mm gauge length extensometer was clamped at the midspan of the tensile dumbbell shape specimen. The tensile test was performed in accordance with ASTM D638 (1991): Standard test method for tensile properties of plastics.

Compression Test
The compressive strength of the adhesives was determined using an Instron 1185 Universal Testing Instrument equipped with a 100 kN load cell by driving the crosshead at 1 mm/min. The compression test was conducted in accordance with ASTM D 695 test method for compressive properties of rigid plastic.

Thermal Analysis of the Adhesive
The dynamic mechanical properties of the three types of adhesives were measured using a Tritec 2000 DMTA analyzer.

Water Absorption Test
Plates of adhesive, 2 mm thick, 500 mm wide and 500 mm long were prepared in a mould from the CB10TSS, timberset and albipox adhesives. Each adhesive was left in the mould for 10 days to cure. After demoulding, the adhesive plate was cut using a

diamond cutter to prepare the moisture uptake test specimens in accordance with BS EN 2243-5:1992: Standard test methods for structural adhesives – Part 5: Ageing tests.

Microscopic Examination

The fracture surface of adhesives and clay particles (quartz, bentonite, mica) were inspected in a JEOL JSN6310 SEM equipped with a computer image analysis system, after gold coating. The TEM measurements were carried out with a 1200 EX transmission electron microscope applying an accelerating voltage of 120 keV.

4.3 DISCUSSIONS AND RESULTS

4.3.1 Effect of Curing Time on Bending Properties of Epoxy-based Adhesives

There is a need to monitor the evolving properties of the adhesive material during curing for structural safety reasons. The behavior of the adhesive during cure can be studied by many methods including x-ray diffraction [33], ultrasonic measurements [34] and differential scanning calorimetry [35]. These methods give an insight into development of the polymer structure during cure and can therefore be used to monitor the cure process. Of considerable importance is the mechanical performance of the adhesive and there are many mechanical tests that can be made on cured samples. This study measured the flexural strength of the adhesive cured at different curing times namely 7, 10, 15, 20, 30, and 40 days to investigate the change in strength at those curing times. Figure 1 shows the effect of curing time on bending strength and MOE.

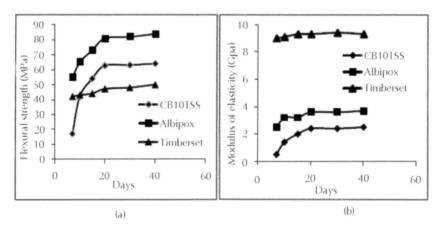

(a)   (b)

FIGURE 1   The effect of curing time on: (a) bending strength and (b) MOE.

Figure 1(a) shows that except for timberset, in the period from 7 to 20 days, the strength of the adhesives are still developing, which indicates that cross-linking is still in progress. The curing of timberset is completed by 7 days or maybe earlier as testing was only started after 7 days. This figure also shows that the curing process is not stopped once the plateau-like region is reached but only slows down when full cross-linking has occurred.

Figure 1(b) shows that the adhesive also becomes stiffer as the process contin- ues. Figure 1(a) suggests that CB10TSS and Albipox required at least 20 days before reaching a fully cured state. This may have implications for the testing of adhesive joints cured at room temperature for less than 20 days. Great care should be taken in industrial applications where the adhesive is cured at room temperature. Based on these results, further mechanical testing was only conducted after 20 days.

4.3.2   The Effect of Fillers on Mechanical Properties of Epoxy-based Adhesive

Figure 2 shows typical stress-strain behavior from tensile tests. All specimens failed immediately after the tensile stress reached the maximum value, however, the stress- strain curves showed considerable nonlinearity before reaching the maximum stress, but no obvious yield point was found in the curve. The plots clearly show enhanced tensile properties for filled epoxy both in terms of stress at break and MOE.

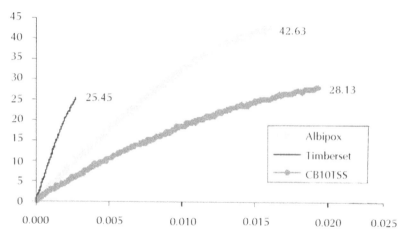

FIGURE 2   Typical examples of stress-strain diagram of epoxy-based adhesive stresses in tension.

Table 1 summarizes the mechanical properties for the entire adhesive used in this study. As seen from Table 1, Albipox has the highest bending, tensile and compression strength followed by CB10TSS and timberset.

TABLE 1   Summary of the mechanical properties of the adhesives.

Sample code		Tensile Properties		Bending Properties		Compression Properties	Strain at break (%)
		Strength (MPa)	MOE (GPa)	Strength (MPa)	MOE (GPa)	Strength (MPa)	
CB10TSS	Mean	26.8	2.4	62.6	2.4	76.3	1.93
	COV (%)	14.14	19.58	8.67	5.37	9.12	

TABLE 1   *(Continued)*

Sample code		Tensile Properties		Bending Properties		Compression Properties	Strain at break
		Strength (MPa)	MOE (GPa)	Strength (MPa)	MOE (GPa)	Strength (MPa)	(%)
Albipox	Mean	42.2	3.8	80.9	3.6	76.5	1.59
	COV(%)	5.38	6.84	5.79	4.99	7.90	
Timberset	Mean	24.1	11.4	47.1	9.3	65.4	0.27
	COV(%)	5.27	12.23	5.80	10.29	4.5	

The inclusion of rubber has improved the Albipox bending and tensile strength by 29 and 57% compared to standard adhesive, CB10TSS. An increase in Albipox strength can also mean that the stress has been transferred from the epoxy matrix to the rubber particles. There is also lower variability observed in the bending and tensile strength of Albipox, which may indicate the homogeneity of the composite. However, the bending strength is almost double the tensile strength. This is due to the difference in the deformation mode where only the outer surfaces of the beam experience the peak stresses and strains. The mean strength of Albipox is at the upper bound of values for commercial epoxy systems [36]. Timberset composites containing ceramic particles showed much lower bending and tensile strength than the neat adhesive, CB10TSS by a factor of 25%. In spite of the reduction in tensile, bending and compression strength for timberset caused by the filler particles, the MOE for Timberset is seen to be more than three times greater than CB10TSS.

From an engineering point of view, elongation at break is an important parameter for describing the rupture behavior of composite materials. Fillers with higher stiffness than the matrix can increase the modulus of the composites but generally fillers cause a dramatic decrease in the elongation at break unless there is good adhesion between the filler and the matrix. Table 1 clearly indicates this is also the case when nanorubber, quartz, mica, and bentonite are used. This demonstrates that the addition of nanofillers is not able to increase elongation at break.

For Albipox, even though has shown lower elongation at break but the bending and tensile strength is significantly improved. This is due to higher energy being absorbed in order to encourage excessive shear yielding which can be seen as white bands in the SEM image in Figure 3(a) taken from the fracture surface of tensile specimen. The white bands are the result of the contrast factor in the secondary electron image (SEI) indicating uneven ridges formed by plastic deformation. At higher magnification, small holes appear in the Albipox adhesive (Figures 3(b) and 3(c)). The generation of these holes is due to cavitations as rubbery particles are pulled out in response to the dilatational stress field, as a crack propagates through the material. This is an important energy dissipating mechanism in rubber toughened epoxies. However, a more important factor is that such voids greatly enhance shear yielding [13].

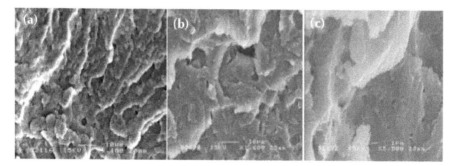

FIGURE 3   The SEM images of tensile fracture surface of Albipox: (a) (1,400x), (b) (1,600x), and (c) showing holes in the adhesive (5,500x).

The SEM images were also taken on the failed bending specimen at the compression and tension edges and are presented in Figure 4. Vertical fissures were found at the compression edge (Figure 4(a)) directly under the loading point but plastic deformation in the form of horizontal yield zones was observed at the tension edge (Figure 4(b) and 4(c)). These extensive overlapping regions of local shear deformation explain why rubbery particles are effective in increasing the tensile and bending strength.

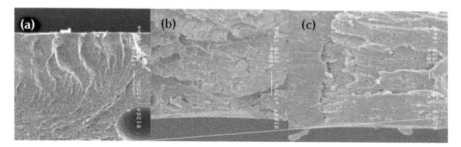

FIGURE 4   The SEM micrographs of fracture surfaces of Albipox, (a) at the compression edge (x250), (b) at tension edge (x230), and (c) showing shear and plastic deformation at the tension edge (x550).

The plastically deformed fracture surfaces of CB10TSS (Figure 5(a)) shows less shear yielding which resulted in lower tensile strength compared to Albipox. Timberset exhibits a planar fracture surface (Figure 5(b)) which indicates brittle fracture behavior and this result in a lower bending and tensile strength for timberset compared to the other adhesives under consideration. The ceramic filler particles appear to have debonded from the matrix (Figure 5(c)), which suggests the presence of weak interfaces between the particles and matrix and according to Young this can lead to a larger inherent flaw size and lower fracture strength. The inhomogeneous interface and poor adhesion may affect the ability to absorb more energy during testing and to sustain the ductility of the matrix. Besides that, the quartz and mica particles are more brittle than the adhesive and dominate the tensile behavior resulting in small strain to failure. Therefore, the increase in strength depends on the strength of filler/adhesive bonding while the elongation at break depends on the extensionality of the fillers and adhesives.

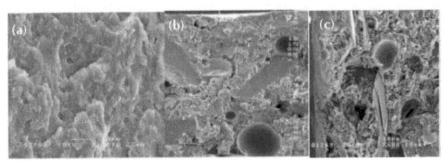

FIGURE 5   Scanning electron micrograph of fracture surfaces following tensile testing of: (a) CB10TSS, x1, 600, (b) Timberset, x220 and (c) Timberset, x600.

To explain the contribution of nanorubber in enhancing the properties of the Albipox, further investigation was made on molecular structure of the matrix using transmission electron micrograph TEM as shown in Figure 6.

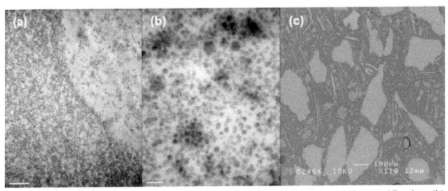

FIGURE 6   Micrographs: (a) TEM micrograph of Albipox adhesive at x15k magnification, (b) TEM micrograph of CB10TSS at higher magnification (x100k) (bar scale dimension = 100 nm), and (c) SEM back scattered signal analysis on timberset.

The first important requirement of rubber in the matrix is to form a separate phase when the resin is cured. If the rubber is too compatible with the resin no phase separation or limited phase separation will occur. This may lead to a small increase in the toughness but to a major decrease in mechanical strength [15]. Figure 6(a) confirmed the phase separation because the boundaries of the dispersed particles appear sharp resulting in increase in tensile, bending and compressive strength. In Figure 6(b) the silica fume in CB10TSS shows an obvious segregation which means that the silica particles are uniformly distributed. From SEM micrographs on timberset (see Figure 6(c)), the particles were seen to contain silicon, iron, sodium, and potassium and these high modulus particles have caused the increase in the MOE.

Figure 7 shows the different modes of failure observed during compressive test. Two distinct fracture modes can be seen. Timberset specimens were found to fail in shear and this diagonal splitting is due to the higher density of the ceramic filler additions which did not allow them to be compressed. In contrast, CB10TSS and Albipox failed due to crushing with a barrel-shaped deformation.

FIGURE 7 Modes of failure of the adhesives after compressive testing, (a) Timberset failed in shear, (b) untested CB10TSS sample, and (c) CB10TSS sample with permanent (barrel-shaped) plastic deformation.

4.3.3 The Effect of Fillers on Thermal Properties of Epoxy-based Adhesive

The effect of filler additions on the dynamic properties of CB10TSS is now compared. Figure 8 shows the plots of storage modulus E′ vs temperature for the different adhesive systems. It is obvious that all of the networks investigated displayed broadly similar dynamic mechanical behavior because the base adhesive is the same. The differences are in the low temperature modulus and the temperature of the main transition from elastic to viscoelastic behavior. Such differences are related to different filler types.

Reinforcement of CB10TSS with phase separated nanorubber particles CTBN remarkably improves glass transition temperature of the standard adhesive (Tg CB10TSS = 31.7°C, Tg Albipox = 42.8°C and Tg Timberset = 53.8°C). The addition of only 2.0% CTBN raised the Tg of CB10TSS by 11°C. The same trend is also observed for timberset. The addition of microfillers to CB10TSS improves the Tg by 22°C. Therefore, the reinforcement with nanorubber and microsized ceramic particles satisfied one of the key aims of this study, to raise the glass transition temperature of CB10TSS. This research has deduced that the higher glass transition temperatures of Albipox and timberset are related to their high cross-link density.

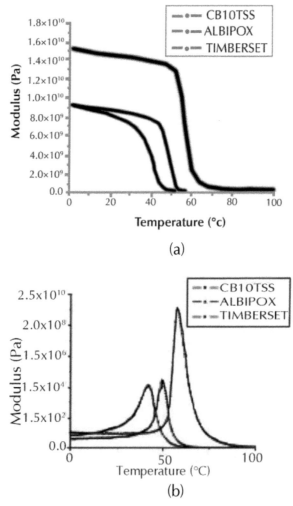

(a)

(b)

FIGURE 8   (Continued)

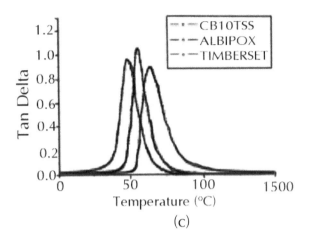

FIGURE 8   Thermal properties of all adhesives: (a) Storage modulus *vs* temperature, (b) Loss Modulus *vs* temperature, and (c) Tan delta *vs* temperature, at 1 Hz.

The E′ curves show that the addition of filler gives higher E′ values compared to standard adhesive (CB10TSS). The ceramic particles in timberset stiffen the adhesive (Figure 6(a)), giving the highest value of E′. This observation agrees with the results for the static elastic modulus found earlier in tensile test. At 20°C, the storage modulus of the adhesives decreases in the order of Timberset > Albipox > CB10TSS. The dynamic modulus was found to be higher than static modulus by a factor of 0.23–3 times higher. The rubbery plateau modulus for timberset is found to be higher than the other adhesives. The addition of high modulus ceramics particle has allowed timberset to retain a higher modulus even at a temperature above the glass transition. The modulus in the rubbery plateau region is proportional to either the number of cross-links or the chain length between entanglements [37]. For the three adhesives other than timberset the storage moduli are similar and very small in the rubbery zone.

The loss modulus, E″, is related to the amount of energy lost due to viscous deformation. The E″ curves for the adhesives are shown in Figure 6(b). The addition of high modulus ceramic particle has allowed timberset to retain a higher modulus even at a temperature above the glass transition. The loss modulus is also increased in the same order as storage modulus. Timberset is again seen to have the highest E″ peak value at 58.13°C (Figure 6(b)) which means there is large amount of damping in the material as a result of the high fraction of filler particles. According to Barral et al. [38], small loss factor peaks can be explained by a relaxation in which the epoxy network losses mobility and free volume as it cools down towards the equilibrium glassy state and, as a result, the ability to dissipate energy is reduced. Results from several experiments support this conclusion such as: from viscosity tests on rate of cross-linking at approximately the same shear rate, CB10TSS has a lower viscosity than Albipox and CB10TSS requires less time to reach a specific value of viscosity compared to Albipox [39]. From the graph of flexural strength *vs* curing time (Figure 1), Timberset reached a stable flexural strength after 7 days and Albipox after 20 days. However, the flexural strength of CB10TSS was still developing after 20 days and was about to reach a steady state of flexural strength after 30 days. Hence timberset has a higher cross-link density followed by Albipox and CB10TSS (The density of Timberset > Albipox > CB10TSS).

The tan δ peak of Albipox is seen to be much higher. This is probably due to better interfacial effects as observed in the SEM micrographs (Figure 3). The tan δ peak for timberset was found to be lower and broader than the other adhesives. The lower tan δ peak reflects poor adhesion between quartz, mica, and bentonite particles and the adhesive matrix (see Figure 5(c)) and the broader peak may indicates higher cross-linking reflecting lower molecular chain mobility.

4.3.4   The Effect of Fillers on Water Absorption Characteristics of Epoxy-based Adhesive

The water absorption was determined by gravimetric measurements given by Equation (1).

$$M_t = \frac{m_t - m_o}{m_o} \times 100\% \tag{1}$$

where M_t is moisture uptake at any time t, m_t is the mass of the specimens at any time t during ageing and m_o is the oven dry mass of the specimen.

Results from the absorption measurements were plotted as moisture uptake, M_t *vs* the square root of time ($t^{1/2}$). Figure 8 shows the absorption curves for CB10TSS, Albipox and Timberset after soaking in water at room temperature for 6 month.

Each point on the curves represents the average of three specimens. The experiments were stopped when the specimens reached the equilibrium state. The absorption of moisture was not accompanied by any visible damage to the material. It can be seen that follow Fickian diffusion behavior with the M_t increasing rapidly in proportion to $\sqrt{t}$ and then the rate increases slowly until saturation. In general, CB10TSS absorbs more moisture than Albipox and Timberset absorbs the least amount of water. Therefore, the addition of CTBN in Albipox and ceramic particles in timberset helps in reducing the moisture absorption.

In the earlier work on mechanical testing and thermal analysis of the adhesives, Timberset was found to have a higher cross-link density. Thus, the high cross-link density of the network minimizes the availability of molecular sized holes in the polymer structure. On the other hand, CB10TSS has the lowest cross-link density compared to Albipox and Timberset. The high moisture absorption is probably due to the looser polymer network which means that it is much easier for moisture to get into intermolecular spaces.

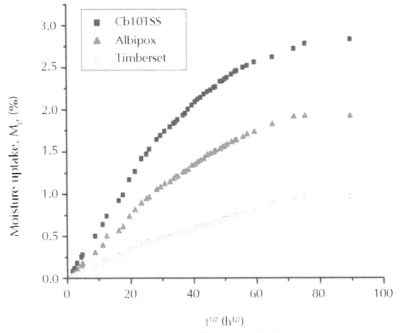

FIGURE 9   Moisture absorption curves for adhesives soaked in water.

4.4   CONCLUSION

The mechanical and thermal properties of nanofilled adhesives have been investigated with the following conclusions:

(1)   CB10TSS adhesive was modified by incorporating fillers namely nanorubber and microsized ceramics particles with the aim to improve mechanical properties and glass transition temperature. Nanorubber addition causes the greatest improvement in strength.

(2)   The full strength of the room temperature cured epoxy adhesives CB10TSS and Albipox is achieved after approximately 20 days whereas timberset is fully cured after 7 days.

(3)   CB10TSS shows higher ductility compared to the filled adhesives Albipox and Timberset but the addition of fillers improves the flexural, tensile, compression, and MOE.

(4)   The SEM provides evidence of the quality of on the interfacial bonding between the matrix and the filler. The coarser surface morphology in Albipox compared with CB10TSS indicates the enhancement in bonding strength which allowed for the stretching and yielding. However, Timberset shows weak bonding which results in a brittle failure mode.

(5)   Timberset possesses the highest storage modulus (E'), loss modulus (E'') and glass transition temperature due to the addition of high modulus ceramic par-

ticles and lowest tan δ value is due to poor interfacial bonding and this corresponds well with measured flexural properties and SEM observations.

(6) The inclusion of fillers improves the T_g significantly by up to 22°C. Albipox has the best combination of properties with maximum flexural strength of 81 MPa, tensile strength of 42 MPa, and Tg of 43°C, compared to 63 MPa, 27 MPa, and 32°C respectively for CB10TSS.

(7) The addition of CTBN (Albipox) and ceramic particles (Timberset) reduces moisture uptake due to higher cross-link density.

KEYWORDS

- Nanoparticles
- *In situ* timber bonding
- Epoxy adhesive
- Glass transition temperature
- Environmental stability

ACKNOWLEDGMENT

The work reported here was financially supported by the E-science fund from Ministry of Science, Technology and Innovation Malaysian. The advice and support by Rotafix Ltd is gratefully acknowledged. We wish to thank technical staff in the materials engineering group at the Universiti Teknologi Mara for their assistance and support.

REFERENCES

1. Alam, P., Ansell, M. P., and Smedley, D. *Flexural properties of steel and FRP reinforced LVL composites*. In *Proceedings of the 8th World Conference on Timber Engineering (WCTE 2004)*. III, Lahti, Finland 342–346 (2004).
2. Davis, M. and Bond, D. Principles and practices of adhesive bonded structural joints and repairs. *Int. J. Adhesion and Adhesives*, **19**, 91–105 (1999).
3. Bainbridge, R. J. and Mettem, C. J. A Review of Memont-Resistant Structural Timber Connections. Institute of Civil Engineers. *Structures and Buildings*, **128**(4) (1998).
4. Wheeler, A. S. and Hutchinson, A. R. Resin repairs to timber structures. *Int. J. Adhesion and Adhesives*, **18**(1), 1–13 (1998).
5. Comyn, J. *Polymer Permeability*. Elsevier Applied Science Publishers, London and New York, p. 177 (1985).
6. Kinloch, A. J. *Adhesion and Adhesives: Science and Technology*. Chapman and Hall, New York, pp. 188–205 (1987).
7. Mays, G. and Hutchinson A. R. *Adhesive in Civil Engineering*. Cambridge University Press, Cambridge (1992).
8. Joseph, D. R. Flitched beams for use in domestic flooring. Final Year Research Project. *Faculty of the Built Environment*, University of the West of England, UK (1999).
9. Davis, T. J. and Claisse, P. A. Resin-injected dowel joints in glulam and structural timber composites. *Construction Building Material*, **15**, 157–167 (2001).
10. Broughton, J. G. and Hutchinson, A. R. Efficient timber connection using bonded-in GFRP rods, composite construction. In *Proceeding of International Conference*. J. Figueiras, et al (Eds.). On Composites in Construction, Balkema, pp. 275–280 (2001).

11. Harvey, K. and Ansell, M. P. Improved timber connections using bonded-in GFRP rods. In *Proceedings of 6th World Conference on Timber Engineering*. Whitsler, British Columbia Paper P04 (2000).
12. Carpenter, W. C. A comparison of numerous lap joint theories for adhesively bonded joints. *Journal of Adhesion*, **35**, 55–73 (1991).
13. Kinloch, A. J. and Young, R. J. *Fracture behavior of polymers*. Applied Science, London, pp. 421–471 (1983).
14. Davis, G. The performance of adhesive systems for structural timbers. *International Journal of Adhesion and Adhesives*, **17**, 247–255(1997).
15. Kinloch, A. J. *Rubber-toughened thermosetting polymers in Structural Adhesive*. A. J Kinloch (Ed.). Elsevier Applied Science, England (1986).
16. Kunz, S. C., Sayre, J. A., and Assink, R. A. Morphology and toughness characterization of epoxy resins modified with amine and carboxyl terminated rubbers. *Polymer*, **23**(13), 1897–1906 (1982).
17. Kinloch, A. J., Shaw, S. J., Tod, D. A., and Hunston, D. L. Deformation and fracture behavior of a rubber-toughened epoxy I Microstructure and fracture studies. *Polymer*, **24**, 1341–54 (1983).
18. McGarry, F. J. *Building Design with Fibre Reinforced Materials*. In Roy. Soc. London, A319, pp. 1536–59 (1970).
19. Ratna, D. and Banthia, A. K. Toughened epoxy adhesives modified with acrylate based liquid rubber. *Polymer Int.*, **49**(3), 309–318 (2000).
20. Zhang, M. Q., Rong, M. Z., Zhang, H. B., and Friedrich, K. Mechanical properties of low nano-silica filled high-density polyethylene composites. *Polymer Eng. Sci.*, **43**, 490–500 (2003).
21. Meguid, S. A. and Sun, Y. On the tensile and shear strength of nano-reinforced composite interfaces. *Materials & Design*, **25**(4), 289–296 (2004).
22. Preghenella, M., Pegoretti, A., and Migliaresi C. Thermo-mechanical characterization of fumed silica-epoxy nanocomposites. *Polymer*, **46**(26), 12065–12072 (2005).
23. Ragosta, G., Abbate, M., Musto, P., Scarinzi, G., and Mascia, L. Epoxy-silica particulate nanocomposites Chemical interactions, reinforcement and fracture toughness. *Polymer*, **46**(23), 10506–10522 (2005).
24. Gledhill, R. A. and Kinloch, A. J. Environmental failure of structural adhesive joints. *Journal of Adhesion*, **6**(4), 315–330 (1974).
25. Ivanova, K. I., Pethrick, R. A., and Affrossman, S. A. Investigation of hydrothermal ageing of a filled rubber toughened epoxy resin using DMTA and dielectric spectroscopy. *Polymer*, **41**, 6787–6796 (2000).
26. Apicella, A., Nicolais, G., Astarita, G., and Drioli, E. Effect of thermal history on water sorption, elastic properties and glass transition of epoxy resins. *Polymer*, **20**, 1143–1148 (1979).
27. Diamant, Y., Marom, G., and Broutman, L. J. The effect of network structure on moisture absorption of epoxy resins. *Journal of Applied Polymer Science*, **26**, 3015–3025 (1981).
28. Smith, L. S. and Schmitz, V. The effect of glass transition temperature of Poly (methyl methacrylate). *Polymer*, **29**, 1871–1878 (1988).
29. Antoon, M. K. and Koening, J. L. Irreversible effects of moisture on the epoxy matrix in glass-reinforced composites. *Journal of Polymer science Polymer Physics Edition*, **19**, 197–212 (1981).
30. Loh, W. K., Crocombe, A. D., Abdel Wahab, M. M., and Ashcroft, I. A. Modeling anomalous moisture uptake, swelling and thermal characteristics of a rubber toughened epoxy adhesive. *International Journal of Adhesion and Adhesives*, **25**, 1–12 (2005).
31. Kinloch, A. J. Interfacial fracture mechanical aspects of adhesive bonded joints Review. *Journal of Adhesion*, **10**(3), 193–219 (1979).
32. Brewis, D. M., Comyn, J., Raval, A. K., and Kinloch A. J. The effect of humidity on the durability of aluminium-epoxide joints. *Int. J. Adhesion and Adhesive*, **10**(4), 247–253 (1990).
33. Lovell, R. and Windle, A. H. A WAXS investigation of local-structure in epoxy networks. *Polymer*, **31**, 593–601 (1990).
34. McCrum, N. G., Read, B. E., and Williams, G. *Anelastic and dielectric effects in polymeric solids*. Dover, New York (1991).

35. Kay, R. and Westwood, A. R. DSC investigation on condensation polymer I Analysis of the curing process. *European Polymer Journal*, **11**(1), 25–30 (1975).
36. Ashby, M. F. *Material selection in mechanical design.* Butterworth-Heinemann, 2nd ed., England (1999).
37. Menard, K. P. *Dynamic Mechanical Analysis A Practical Introduction.* CRC Boca Roton, Florida (1999).
38. Barral, L., Cano, J., Lopez, J., Lopez-Bueno, I., Nogueira, P., Ramirez, C., Torres, A., and Abad, M. J. Thermal properties of amine cured diglycidyl ether of bisphenol a epoxy blended with poly(ether imide). *Thermochimica Acta*, **344**(1–2), 137–143 (2000).
39. Ahmad, Z., Ansell, M. P., and Smedley, D. Thermal properties of epoxy-based adhesive reinforced with nano and micro-particles for in-situ timber bonding. *Int. J. of Eng. and Tech.*, **10**(2), 32–38 (2010).

5 Magnetic Ferro-photo Gels for Synergistic Reduction of Hexavalent Chromium in Wastewater

Ani Idris, Nursia Hassan, and Audrey-Flore Ngomsik

CONTENTS

5.1   INTRODUCTION

Photocatalysis has the capability to destroy pollutants or convert them into less toxic substances, in contrast to current treatment methods such as adsorption where contaminants are merely concentrated by transferring them onto the adsorbent but they do not convert into non toxic wastes [1]. It should be noted that, for photocatalytic reduction process, two methods of catalyst metal oxide application are preferred: (i) suspension in aqueous media [2] and (ii) immobilization on suitable support material [3]. Although suspended photocatalyst systems always give higher reduction rates due to the larger surface area when compared to the immobilized system, there is one obvious problem arising from it. The nanosized particles in the range of 30–200 nm [4] needs to undergo separation process by which the system must be installed with a liquid-solid separator, which is expensive and constitutes a major drawback in the commercialization of this system. The second disadvantage of the freely suspension system is that nanometer particles from the effluent may cause turbidity in the downstream. Thus, there is a need to develop a practical technology for the separation of the catalyst and effluent.

The search for photocatalysts with enhanced separation properties has resulted in use of maghemite nanoparticles. Recently, [5] have developed photocatalyst magnetic separable beads for the Cr(VI) reduction under sunlight and the results revealed that the photo reduction of Cr(VI) was completed in just after 50 min. However, the experiments were performed using synthetic Cr(VI) solutions. Thus, this study addresses the application of photocatalyst magnetic separable beads in treating wastewater with high concentration of carcinogenic hexavalent chromium (390 mg/l) for slurry type photo reactors.

5.2 EXPERIMENTAL

5.2.1 Preparation of Magnetic Ferro-photo Gel

Magnetic ferro-photo gels containing magnetic nanosized iron oxide in the range of (5 nm < d < 20 nm) coated with a biopolymer extracted from algae, sodium alginate was prepared according to the method described by [5]. The magnetic material used is a ferrofluid composed of maghemite nanoparticles dispersed in an aqueous solution. They were synthesized according to the process described by [6] and improved by [7]. This method allows the size and the superficial charge of particles to be controlled.

5.2.2 Electroplating Wastewater

The actual industrial wastewater was sampled from electroplating company in Masai, Johor and kept in plastic container at room temperature [8]. The photocatalytic activity of the magnetic beads was tested for Cr(VI) wastewater in acidic pH. The magnetic beads with the 16% (v/v) photocatalyst dosage were used throughout the experiment. However, the amounts of beads used were varied at 20 and 30 g to treat 100 ml of Cr(VI) wastewater. The solutions were exposed to sunlight and all the experiments were performed in triplicate during the sunny days, from 10.00 a.m. to 14.00 p.m. using DPC method.

5.2.3 Analytical Method

The reduction of chromium (VI) was determined colorimetrically at 540 nm using the diphenylcarbazide (DPC) method (UV–vis spectrophotometer UV-1240, SHIMADZU) with a detection limit of $5\mu g l^{-1}$. The red-violet to purple color formed was then measured at OD_{540} using distilled water as reference [9].

5.3 DISCUSSION AND RESULTS

The overall process procedure of magnetic ferro-photo gels as well as their photocatalytic reduction of the chromium in wastewater is illustrated in Figure 1. Results from the characterization of the wastewater are as follows: color, yellow to dark orange; pH, between 2.10 and 2.30 ± 0.05; temperature, 30.7 ± 3.39°C. The initial Cr(VI) content as determined by DPC method was found to be around 390 mg/l which exceeded the prescribed legal limit.

Figure 2 shows the result of the photoreduction of hexavalent chromium in wastewater and the reduction was monitored as a function of time and amount of beads loading. As the amount of beads is increased from 20 to 30 g the degradation efficiency of Cr(VI) is enhanced, and reached the maximum of 100% after 2 hr treatment, when the amount of beads was 30 g.

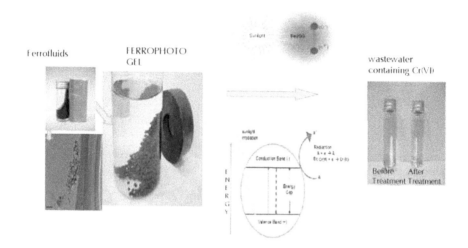

FIGURE 1   The fabrication procedures of magnetic ferro-photo gels and their photocatalytic reaction.

The rate initially increases with an increase in the catalyst loading, as this translates into an increase in the number of available active sites for adsorption and hence photoreduction [2]. In this case, the magnetic ferro-photo gels which consist of γ-Fe$_2$O$_3$ which produces electron hole pair is responsible for the Cr(VI) reduction to Cr(III). Under sunlight irradiation, electron (e$^-$), and hole (h$^+$) pairs were produced and the 3e reaction of Cr(VI) is reduced to Cr(III), while the conjugate anodic reaction was the oxidation of water to oxygen demonstrated by Equations (1)–(3).

Cr(VI) reduction probably occurred *via* three subsequent 1e transfer processes, ending in Cr(III) which is the stable final product.

$$Cr(VI)^{e-} \rightarrow Cr(V)^{e-} \rightarrow Cr(IV)^{e-} \rightarrow Cr(III) \tag{1}$$

$$2H_2O + 4H^+ \rightarrow O_2 + 4H^+ \tag{2}$$

The overall reaction at extreme acidic conditions (pH 1 and 2) are as follows:

$$Cr_2O_7^{2-} + 14H^+ + 6e^- \rightarrow 2Cr^{3+} + 7H_2O \tag{3}$$

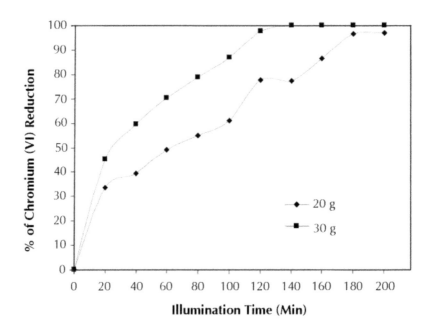

FIGURE 2   Influence of amount of beads on photocatalytic reduction of Cr(VI).

5.4   CONCLUSION

In short, the present work illustrated the use of magnetic ferro-photo gels in treating effluents contaminated with Cr(VI). This new approach might be applicable as an alternative or complementary method for the treatment of inorganic pollutants in wastewater as it is a sunlight driven photocatalyst. This enables the development of a commercially viable photocatalytic water treatment processes which directly compete with currently available technologies for inorganic removal from wastewaters. In addition, these magnetic ferro-photo gels allow for easy recovery from the treated water by magnetic force, without the need for further downstream treatment processes. Once separated, this photocatalyst can be reused because of its regenerative property under photocatalytic reaction. Finally, these magnetic ferro-photo gels contribute towards green and sustainable technology.

KEYWORDS

- **Diphenylcarbazide method**
- **Immobilized system**
- **Magnetic ferro-photo gel**
- **Photocatalysis**
- **Wastewater**

REFERENCES

1. Beydoun, D., Amal, R., Low, G., and McEvoy, S. Role of Nanoparticles in Photocatalysis. *Journal of Nanoparticle Research*, **1**, 439–458 (1999).
2. Mohapatra, P., Samantaray, S. K., and Parida, K. Photocatalytic Reduction of Hexavalent Chromium in Aqueous Solution over Sulphate Modified Titania. *Journal of Photochemistry and Photobiology A Chemistry*, **170**, 189–194 (2005).
3. Tuprakay, S. and Liengcharernsit, W. Lifetime and Regeneration of Immobilized Titania for Photocatalytic Removal of Aqueous Hexavalent Chromium. *Journal of Hazardous Material*, **B124**, 53–58 (2005).
4. Senthilkumaar, S., Porkodi, K., Gomathi, R., Geetha Maheswari, A., and Manonmani, N. Sol-gel Derived Silver Doped Nanocrystalline Titania Catalyzed Photodegradation of Methylene Blue from Aqueous Solution. *Dyes Pigment*, **69**, 22–25 (2006).
5. Idris, A., Hassan, N., Mohd Ismail, N, Misran, E., Yusuf, N., Ngomsik, A. F., and Bee, A. Photocatalytic Magnetic Separable Beads for Chromium (VI) reduction. *Water Research*, **44**, 1683–1688 (2010).
6. Massart, R. Preparation of Aqueous Magnetic Liquids in Alkaline and Acidic Media. *IEEE Transactions on Magnetics*, **17**, 1247–1248 (1981).
7. Bee, A., Massart, R., and Neveu, S. Synthesis of Very Fine Maghemite Particles. *Journal of Magnetism and Magnetic Materials*, **149**, 6–9 (1995).
8. Kiptoo, J. K., Ngila, J. C., and Sawula, G. M. Speciation Studies of Nickel and Chromium in Wastewater from an Electroplating Plant. *Talanta*, **64**, 54–59 (2004).
9. Zakaria, Z. A., Zakaria, Z., Surif, S., and Ahmad, W. A. Hexavalent Chromium Reduction by *Acinetobacter haemolyticus* isolated from Heavy-metal Contaminated Wastewater. *Journal of Hazardous Materials*, **146**, 30–38 (2007).

6 Mode-I Fracture of CFRP Composite Bonded Joints: Influence of Surface Treatment and Adhesive Ductility

K. B. Katnam and T. M. Young

CONTENTS

6.1 INTRODUCTION

Composite materials are increasingly being used for structural applications in the aircraft industry, for example, wings and fuselages, because of their superior damage resistance and excellent stiffness to weight ratio [1, 2]. Carbon fiber reinforced polymer

(CFRP) composites with improved mechanical properties allow structural design to displace the more conventional materials, aluminum and titanium alloys, for primary structures [3]. For example, in the design of modern commercial aircraft such as the Boeing 787, advanced composites (50% by weight) are replacing conventional structural materials [4].

To assemble different composite structural components, structural bonding using toughened epoxy adhesives is an attractive alternative for composite materials as it provides many advantages over conventional mechanical joining methods, for example reduces stress concentrations and provides better fatigue resistance [5, 6]. Although structural bonded joints are thoroughly analyzed using analytical and numerical models, to attain a reliable design for bonded structures, a detailed experimental testing is often necessary to examine the failure of bonded structural joints for a given combination of adherend/surface treatment/adhesive [7]. An important design aspect is the fracture behavior of bonded composite joints under mode-I stress conditions [8].

In this work, double cantilever beam (DCB) tests were conducted to determine the mode-I fracture energy of bonded composite joints. The composite adherend used was a symmetric pre-preg laminate (Hexcel HTA/6376). Two composite surface treatments were investigated: (a) methyl-ethyl-ketone wipe and ultrasonic (MEK-U) bath, and (b) low pressure O_2 plasma [9], (for example, for full details on this process). The surface treatments were tested in combination with two research grade epoxy paste adhesives. The fracture energy values for the four adherend surface treatment adhesive combinations were determined and the fracture surfaces were examined using scanning electron microscopy (SEM) to identify failure mechanisms. The experimental test results and fractographic evidence obtained are presented and discussed in this chapter.

Structural bonding using toughened epoxy adhesives is an attractive alternative to conventional mechanical joining techniques for composite structural applications in the aircraft industry, for example wings and fuselages. However, an important design aspect is the fracture behavior in bonded joints under mode-I stress conditions (i.e. peel dominated fracture). In this work, DCB tests were conducted to determine the mode-I fracture energy of CFRP composite bonded joints with two different epoxy adhesives. Two composite surface treatments were investigated: (a) MEK-U bath, and (b) low pressure O_2 plasma. The fracture energy values were experimentally determined, and the obtained fracture surfaces were examined using SEM to identify failure mechanisms. The results indicated that interfacial adhesion was considerably improved with low pressure O_2 plasma treatment. Furthermore, adhesive ductility was observed to have positively influenced the fracture behavior of composite bonded joints.

6.2 MATERIALS AND METHODS

6.2.1 Pre-preg Laminates

Unidirectional pre-preg material, Hexcel HTA6376, was employed to lay-up composite laminates. A laminate thickness of 2 mm was targeted using 16 plies (each ply is approximately 0.125 mm thick) with $(0/90)_{4S}$ symmetric lay-up. The panels were cured at 177°C under 7 bar pressure for 4 hr in an autoclave (TC1000LHTHP system, manufactured by LBBC, UK) using a vacuum bag. A tight woven polyester release ply (peel ply G) was used in the composite curing.

6.2.2 Two-part Epoxy Paste Adhesive

Two research grade structural epoxy paste adhesives, denoted as adhesive X and adhesive Y, were used to bond CFRP composite panels. The two adhesives were supplied in a dual cartridge form and were mixed using a static mixer with a dispenser. The two adhesives were cured at 80°C under 1,000 m bar pressures for 4 hr.

6.3 BONDING AND SPECIMEN PREPARATION

6.3.1 Surface Preparation

The pre-preg composite panels (250 × 250 mm) were rinsed in water to remove the coolant that was used to cut the panels and then dried in oven at 121°C for an hour. The dry surfaces were then treated prior to bonding. In the first case, methyl-ethyl-ketone was used to wipe the dry surfaces and subsequently immersed in an ultrasonic bath (de-ionized water) for 30 min. The panels were then dried in an oven at 121°C for an hour. In the second case, the dry surfaces were exposed to low pressure O_2 plasma (at 500 W, 100 m Torr, and 162 MHz RF) for 30 s. The surfaces were bonded within a 5 hr period after being plasma treated. Water contact angles were measured at different time intervals on the treated surfaces to investigate variation in surface wettability with time. The untreated and treated surfaces gave a contact angle of 78 and 0° degrees, respectively. It was also observed that the water contact angle remained nearly constant (0° degree) within a 12 hr period after being plasma treated.

6.3.2 Bonding in Hot Drape Former

The treated surfaces were bonded by employing a two-part aluminum mould and a hot drape former (HDF2 from laminating technology, UK). Two-part epoxy adhesive (using dual-cartridge and static-mixer in a dispenser) was applied on the treated composite surfaces, and brass shims (0.2 mm) along with a nylon scrim were employed to control the adhesive thickness. Furthermore, a brass shim (50.8 mm wide) with a release film was used to create a pre-crack in the bonded panels. The mould was then placed in the hot drape former and cured at 80°C for 4 hr under 1,000 mbar pressures.

6.3.3 Specimen Preparation

The bonded composite panels were cut using a diamond edged composite cutter to obtain test specimens. To minimize the composite cutting damage a coolant was used and the specimens were immediately wiped with tissue paper and dried at room temperature to reduce moisture diffusion. The dimensional details of the test specimen are given in Figure 1. Metal hinges were bonded to each composite substrate using Scotch Weld 3M 9323 adhesive at room temperature with pressure being applied by G-clamps.

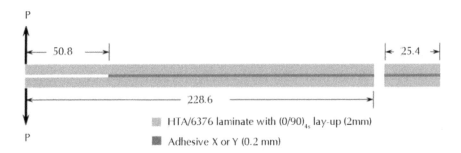

P

|← 50.8 →| |← 25.4 →|

|←——————————————— 228.6 ———————————————→|

P

■ HTA/6376 laminate with (0/90)$_{4s}$ lay-up (2mm)

■ Adhesive X or Y (0.2 mm)

FIGURE 1   The dimensional details of the DCB test specimen.

6.4   EXPERIMENTAL TESTING

6.4.1   DCB Tests

The quasi-static tests were performed under normal conditions at a temperature of 23 ±
2 C and a relative humidity of 50 ± 5%. A tensile testing machine (Tinius Olsen) with a
1,000 N load cell was used. A paper scale was bonded to the edge of the specimen and
an optical microscope was employed to measure the crack length at different displace-
ments. A data-reduction technique recommended in BS 7991 [10] was used to evaluate
the mode-I fracture energy. The load-displacement curves obtained for adhesive X and
adhesive Y for MEK-U and O$_2$ plasma treatments were compared and shown in Figure
2(a) and 2(b), respectively (although a batch of five specimens were tested for each
surface treatment and adhesive combination, representative curves were compared in
Figure 2). In Figure 2(a), a slip-stick fracture pattern was observed (with sharp peaks
in load-displacement curves) for both adhesives with MEK-U treatment. However,
in Figure 2(b), the load-displacement curves obtained for adhesive Y with O$_2$ plasma
treatment showed relatively blunt peaks (for adhesive X the peaks were sharp). This
indicated a higher adhesive ductility during damage initiation in adhesive Y when
compared to adhesive X. The evaluated mode-I fracture energy values were normalized and shown in Figure 3. It was clear that the O$_2$ plasma treatment provided good
interfacial adhesion and considerably improved (a factor of 5.1 for adhesive X and 9.8
for adhesive Y) the mode-I fracture energy values.

6.4.2   Fractography

The fracture surfaces were examined to identify different failure patterns. It was observed that the test specimens with MEK-U treatment failed by interfacial crack ini-
tiation/propagation. However, the test specimens with O$_2$ plasma treatment failed by
crack initiation/propagation either in the bondline or in the composite laminate. The
failure pattern and representative fracture surfaces are shown in Figure 4 for the four
surface treatment and adhesive combinations. From Figures 2(b), 4(c), and 4(d), it
was observed that failure occurred in the composite laminate with some patches of
cohesive failure in the bondline relatively at a lower peak load (when the first peak
loads for adhesive X and adhesive Y were compared) for adhesive X because of its

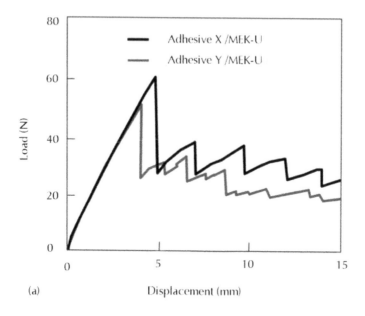

(a)

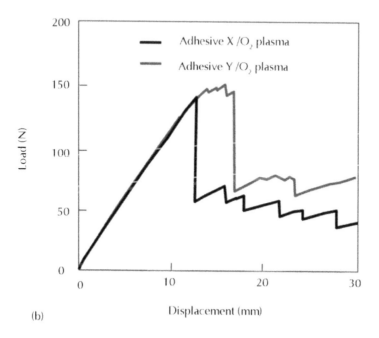

(b)

FIGURE 2   Comparison of load-displacement curves obtained for adhesive X and Y: (a) with methyl-ethyl-ketone and ultrasonic bath (MEK-U) and (b) with low pressure oxygen plasma treatment.

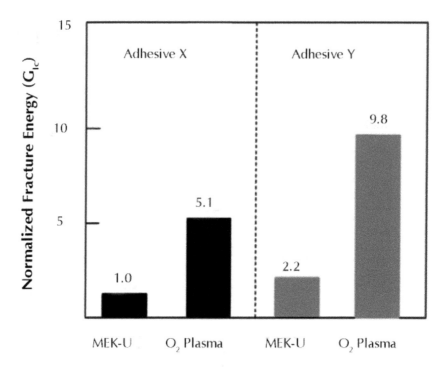

FIGURE 3 Comparison of normalized fracture energy (G_{Ic}) for adhesive X and adhesive Y with methyl-ethyl-ketone and ultrasonic bath (MEK-U) and low pressure O_2 plasma treatments.

brittle behavior. In this regard, the fracture surfaces were further examined using SEM (JEOL JCM 5700) to investigate differences in failure mechanisms in the O_2 plasma treated specimens. The micrographs were taken from cohesive failure regions (near the stick-slip transition region) in the bondline and are shown in Figure 5. The failure surface near the stick-slip transition region for adhesive X is shown in Figure 5(a). Furthermore, the toughening particles used in adhesive X and the slow fracture region is shown in Figure 5(a). Similarly, the failure surface near the stick-slip transition region for adhesive Y is shown in Figure 5(b). The poor adhesion at the interface of the nylon scrim is evident for both adhesives. However, when compared to adhesive X, a considerable difference in surface roughness was observed for adhesive Y indicating higher ductility during damage initiation. This is manifested in the load displacement curves in Figure 3(b), showing a relatively blunt peak for adhesive Y, when compared to the load-displacement curves for adhesive X. Moreover, for O_2 plasma treatment, the normalized fracture energy (G_{Ic}) value for adhesive Y was 9.8, which is considerably higher than the value for adhesive X (only 5.1, as shown in Figure 3). It is thought that the difference in adhesive ductility (relative to adhesive X) during crack initiation might have improved the mode-I fracture energy for adhesive Y by lowering the stress concentration at the crack tip and thus changing the failure locus (i.e. cohesive failure for adhesive Y and composite failure for adhesive X).

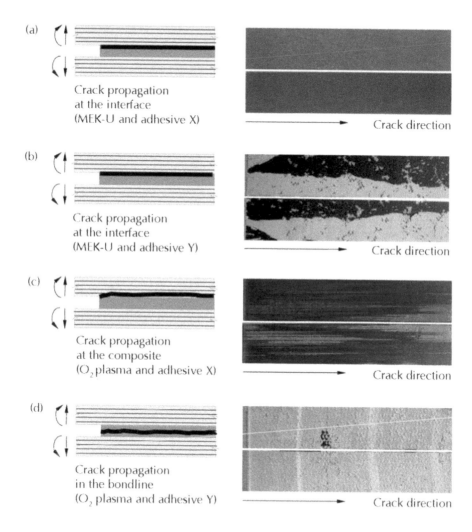

FIGURE 4   Comparison of failure behavior: (a) interfacial failure from MEK-U treatment and adhesive X, (b) interfacial failure from MEK-U treatment and adhesive Y, (c) composite failure from O_2 plasma treatment and adhesive X, and (d) cohesive failure in the bondline (with stress-whitening) from O_2 plasma treatment and adhesive Y.

(a)

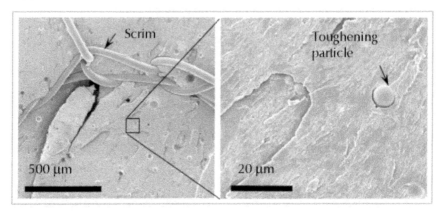

(b)

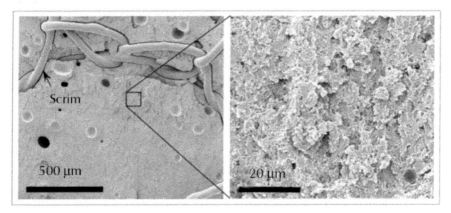

FIGURE 5   Comparison of fracture micrographs obtained for: (a) adhesive X with O$_2$ plasma treatment and (b) adhesive Y with O$_2$ plasma treatments.

6.5   CONCLUSION

The DCB tests were conducted to determine the mode-I fracture energy of CFRP composite bonded joints with two structural adhesives. A pre-preg based laminate was used, and two surface treatments were investigated: (a) MEK-U bath (MEK-U) and (b) low pressure O$_2$ plasma. The fracture energy values were experimentally determined and the obtained fracture surfaces were examined to identify failure mechanisms. The results indicated that interfacial adhesion and mode-I fracture energy were considerably improved with low pressure O$_2$ plasma treatment. Moreover, it was observed that adhesive ductility improved mode-I fracture energy and influenced the locus of failure in the CFRP composite bonded joints.

KEYWORDS

- **Carbon-fiber-reinforced polymer**
- **Composite bonded joints**
- **Mode-I fracture**
- **Plasma treatment**
- **Toughened structural adhesives**

ACKNOWLEDGMENTS

The authors would like to acknowledge: Enterprise Ireland for research funding, Henkel Adhesives (Ireland) and Bombardier Aerospace (Belfast, UK) for providing adhesives, (c) Dr. A. R. Ellingboe (National Centre for Plasma Science and Technology, Dublin City University) and Mr. Clemens Schmidt-Eisenlohr (University of Limerick, Ireland and University of Stuttgart, Germany) for conducting experimental work.

REFERENCES

1. Zagainov, G. I. and Lozino-Lozinski, G. E. *Composite Materials in Aerospace Design*, Chapman & Hall, London (1996).
2. Chung, D. D. L. *Composite Materials Science and Applications*. Springer, London (2010).
3. Soutis, C. *Progress in Aerospace Sciences*, **41**(2), 143–151 (2005).
4. AERO magazine. Boeing, QTR 4, 17–23 (2006).
5. Matthews, F. L., Kilty, P. P. F., and Godwin, E. W. A. *Composites*. **13**, 29–37 (1982).
6. Banea, M. D. and Da Silva, L. F. M. Proceedings of the Institution of Mechanical Engineers, Part L. *Journal of Materials Design and Applications,* **223**, 1–18 (2009).
7. Adams, R. D., Comyn, J., and Wake, W. C. *Structural Adhesive Joints in Engineering*. Chapman & Hall, London (1997).
8. Ashcroft, I. A., Hughes, D. J., and Shaw, S. L. *Int. J. Adhes. Adhes.*, **21**, 87–99 (2001).
9. Shanahan, M. E. R. and Bourges-Monnier, C. *Int. J. Adhes. Adhes.*, **16**, 129–35 (1996).
10. Standard BS 7991. Determination of the mode I adhesive fracture energy (G_{Ic}) of structure adhesives using the DCB and tapered double cantilever beam (TDCB) specimens (2001).

7 Charge Transport Property of Multiferroic BiFeO$_3$ Electro Ceramics

Sangram K. Pradhan, Prajna P. Rout, Sangram K. Das, Viswaranjan Mohanta, and Binod K. Roul

CONTENTS

7.1 INTRODUCTION

Multiferroics is a class of functional materials that simultaneously exhibit at least two or more ferroic behaviors, that is ferroelectric, ferromagnetic, or ferroelastic in a single structure. They have been regarded as ideal candidates for several technologically important applications, owing to the realization of magnetoelectric (ME) effects [1-13]. They can demonstrate not only the ME behavior but also the desired coupling between the magnetic and electric polarizations leading to multifunctionalities such as electric field controlled magnetic data storage or *vice versa* [11]. This unique feature provides an interesting opportunity for exploiting numerous functionalities in a single structure. As part of the technological drive toward device miniaturization, considerable effort has been devoted to the combination of electronic and magnetic properties into one multifunctional material that is a single device component that can perform more than one task [10, 11]. However, BiFeO$_3$ (BFO) is a well studied single-phase multiferroic, which simultaneously exhibits antiferromagnetism and electrical polarizations at room temperature as deserving candidate for practical applications. But due to charge conduction, high leakage current value limits the application of BFO in electronic devices. In this chapter, we describe the detail synthesis procedure as well as variation of metal ion concentration which greatly effect to obtain high purity BFO with rhombohedral phase by conventional solid state reaction route with slow step sintering schedule.

Charge (electrical) and mass (chemical) transport properties of BFO multiferroic and doped BFO electroceramics in its bulk and thin film forms are required to be tailored effectively to suppress the leakage current for real spintronics based functional device applications. Analyzing and understanding the leakage current mechanism that ingrained within the ferroelectric property associated with multifunctional BFO material is infact the basic issue. As known, the major leakage current mechanisms involved in multifunctional electroceramics in its thin film and bulk form are the electrode material interfacial barrier limited Schottky emission, the modified Schottky conduction, the space charge limited bulk conduction and the Poole-Frenkel emission arising from the bulk. Schottky emission is purely an interfacial process phenomenon which normally arises due to barrier formation onto the electrode interface where Poole-Frenkel and space charge limited conduction arise from the bulk of the material. We have prepared Fe-deficient BFO through a novel slow step sintering route and tailored the leakage behavior which promote to achieve appreciable enhanced P-E hysteresis loop. We also observed that Fe deficient BFO helps the formation of Fe^{3+} state which hindered the creation of oxygen vacancies and also favor to reduce this leakage current. Structural distortion of Fe deficient BFO changes Fe-O-Fe bond angle and evolve spin disorder which destruct the spin cycloid and releases locked magnetization that enhanced the ferromagnetism and ferroelectric behavior in Fe deficient BFO multiferroic electro ceramics.

7.2   SAMPLE PREPARATION

High pure (99.999%) Bi_2O_3 and Fe_2O_3 powders were taken with appropriate composition (pure BFO and $Bi_{1.1}Fe_{0.9}O_3$). Powders of respective compositions were heated at 500°C for 6 hr which was subsequently quenched to RT for immediate grinding. This procedure was repeated five times in order to achieve homogeneous mixture with smaller particle size. Powders of respective compositions were pelletized using freshly prepared poly vinyl alcohol (PVA) as binder. Cylindrical pellets having dimension 13 mm diameter and 2 mm thickness were prepared by hydraulic press with a pressure 10 tone/cm^2. Sintering schedule of pellets were carried out at 50°C/hr heating rate with a constant soaking periods of 4 hr at 200 and 600°C. This soaking period is required to remove any volatile impurity/organic materials (binder) from pellets. Finally, samples are sintered at 850°C for 24 hr by slow step sintering schedule. In order to avoid surface contamination, all thick sintered samples were polished with the same powder to obtain thickness down to submillimeter (0.32 mm).

7.3   EXPERIMENTAL

X-ray diffraction pattern of the samples were carried out in the 2θ range (20–80°) using Cu K$_\alpha$ radiation by a Philips diffractometer (Model 1715) fitted with monochromator and operated at 40 KV and 20 mA. The ferroelectric (electric polarization as a function of electric field) measurements of sintered pellets were carried out using ferroelectric hysteresis loop tracer (Radiant technology, USA). Chemical bonding analysis of sample was carried out by using x-ray photoelectron spectroscopy (XPS) (VG-Microtech system with a base pressure of 1×10^{-10}Torr). Current-voltage measurements were carried out using Keithley electrometer (Model 617). Surface contacts for electrical measurements were done by sputtering high pure silver and copper thin electrodes are connected with silver contact using indium. All I-V measurements of the ceramic samples (~0.32 mm thick) were taken at room temperature.

7.4   DISCUSSION AND RESULTS

Figure 1 shows XRD pattern of pure BFO, Bi rich Fe deficient BFO which were sintered at 850°C for 24 hr through slow step sintering schedule. Respective peaks in XRD pattern are indexed with h, k, l indices, and are fitted to R3c space group. However, there is some unidentified h, k, l reflections present in the XRD pattern which are normally found in pure BFO system and are corresponding to $Bi_{36}Fe_2O_{57}$ and $Bi_2Fe_4O_9$ phase. The XRD pattern of a typical Bi rich Fe deficient BFO ($Bi_{1.1}Fe_{0.9}O_3$) is also shown in Figure 1. It is noted that the impurity phases that are observed in pure BFO are completely suppressed in $Bi_{1.1}Fe_{0.9}O_3$. It is inferred from Figure 1 that Bi rich Fe deficient BFO exhibits near monophasic R3c symmetry with impurity free phases with respect to pure BFO.

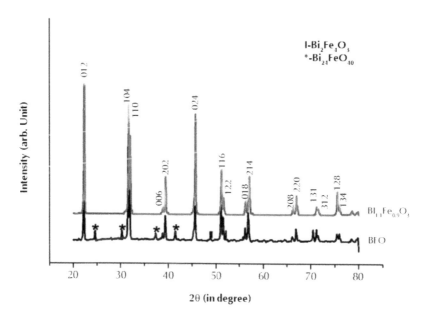

FIGURE 1   The XRD patterns of $BiFeO_3$, and $Bi_{1.1}Fe_{0.9}O_3$ pellets sintered at 850°C for 24 hr using slow step (50°C/hr) sintering schedule.

In order to observe the oxidation state of Fe in $Bi_{1.1}Fe_{0.9}O_3$, chemical bonding analysis by XPS were carried out and a typical spectrum is shown in Figure 2. It is observed from photoemission peaks of $Bi_{1.1}Fe_{0.9}O_3$ sample that peaks at 710.4 and 725.1 eV are corresponding to the spin orbit doublet component of Fe $2p_{3/2}$ and $2p_{1/2}$ in Fe^{+3} oxidation states and no signature of Fe^{2+} state is observed with reduction of leakage current. It is anticipated that Fe deficient may helps to suppress the reduction of Fe from Fe^{3+} to Fe^{2+} state.

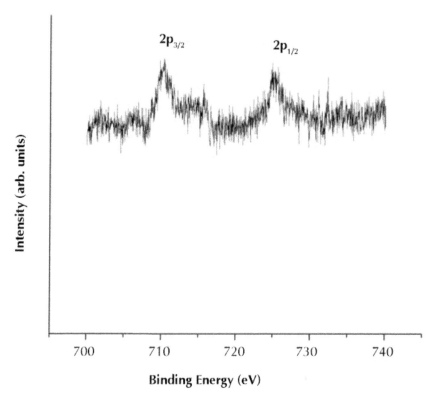

FIGURE 2   The XPS spectra (Fe) of $Bi_{1.2}Fe_{0.8}O_3$ samples sintered at 850°C for 24 hr.

Figure 3 shows the I-V characteristics of BFO and $Bi_{1.1}Fe_{0.9}O_3$ ceramic. Both I-V curves showed good symmetry under the application of positive and negative voltage. However, a significant difference in leakage current is observed in case of $Bi_{1.1}Fe_{0.9}O_3$. At an applied voltage of ± 20 V the leakage current observed in $Bi_{1.1}Fe_{0.9}O_3$ is 3.74 × 10^{-7}A which is about two orders of magnitude lower than pure BFO under application of same applied voltage and shows a significant improvement in lowering of leakage current with varying Bi and Fe concentration. It is known that Bi is highly volatile in nature and it starts melting and slowly evaporates at temperature beyond 820°C and causes non-stochiometric of Bi and Fe in parent BFO compound and causes formation of different types of Bi and Fe rich BFO phase as well vacancies of oxygen and bismuth. As known, the high leakage current in pure BFO ceramics originated mainly from charge defects such as bismuth vacancies $V_{Bi}{}^{3-}$ and oxygen vacancies Vo^{2+}. Vo^{2+} mainly came from Bi volatility and oxygen vacancies from the valance fluctuation of iron from Fe^{3+} to Fe^{2+}. From observation, it is noted that slow step sintering schedule in the air along with deficient of Fe in parent compound of BFO helps to reduce the number of Vo^{2+} and $V_{Bi}{}^{3-}$ and thus to improve the resistivity of $Bi_{1.1}Fe_{0.9}O_3$ samples.

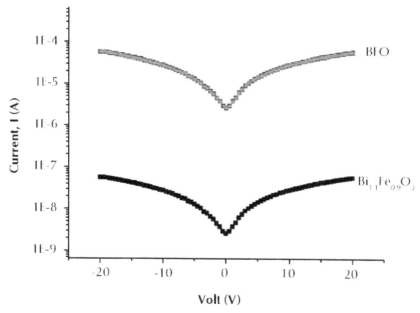

FIGURE 3 Leakage current as a function of forward and reverse bias applied voltage of $Bi_{1.1}Fe_{0.9}O_3$, and $BiFeO_3$ samples sintered at 850°C for 24 hr using slow step sintering schedule.

In order to ascertain the ferroelectric behavior of BFO and $Bi_{1.1}Fe_{0.9}O_3$, polarization as a function of electric field were carried out at room temperature using ferroelectric loop tracer set-up and shown in Figure 4. In case of pure BFO no saturation of P-E curve could be obtained up to the maximum applied of electric field due to leaky nature of the sample but with increase in concentration of Bi and decrease in concentration of Fe ($Bi_{1.1}Fe_{0.9}O_3$), sample shows a good P-E loop having $2P_r$ and $2E_c$ values 8 µ°C/cm^2 and 100 kV/cm respectively by the application of 130 kV/cm applied electric field which is also strongly evidenced from I-V measurement. The P-E hysteresis of pure bismuth ferrite and bismuth rich Fe deficient BFO are shown in Figure 4.

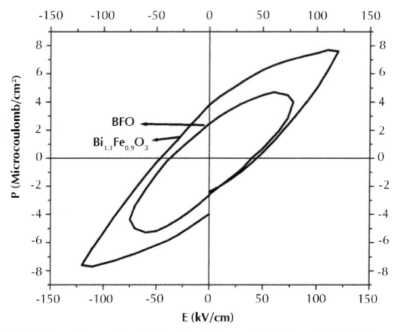

FIGURE 4   Room Temperature P-E hysteresis loops of $BiFeO_3$, and $Bi_{1.1}Fe_{0.9}O_3$ samples sintered at 850°C for 24 hr using slow step sintering schedule.

It is evident that, increase in Bi concentration is likely to compensate Bi loss during sintering at elevated temperature because Bi_2O_3 is highly volatile in nature and its melting point is 817°C and it also causes structural distortion as resulting a well shaped P-E loop is observed in $Bi_{1.1}Fe_{0.9}O_3$.

CONCLUSION

In conclusion, high density $Bi_{1.1}Fe_{0.9}O_3$ ceramics has been synthesized which showed appreciable distortion in the structure and promotes low loss P-E loop as well as suppress the leakage current with respect to original BFO. Fe deficient also helps to suppress the reduction of Fe^{3+} to Fe^{2+} during sintering at 850°C which certainly helps to overcome the long standing leakage current issues associated with lead free BFO multiferroic material.

KEYWORDS

- **Electro ceramics**
- **Magnetoelectric**
- **Poly vinyl alcohol**
- **Schottky emission**
- **X-ray photoelectron spectroscopy**

ACKNOWLEDGMENTS

Author Sangram Keshari Pradhan is gratefully acknowledged the financial support and research facilities received from Institute of Materials Science, Bhubaneswar.

REFERENCES

1. Fiebig, M. J. *Phys. D*, **38**, R123–152 (2005).
2. Zhou, J. P., He, H., Shi, Z., and Nan, C. W. *Appl. Phys. Lett.*, **88**, 013111–13 (2006).
3. Li, Y. W., Sun, J. L., Chen, J., Meng, X. J., and Chu, J. H. *Appl. Phys. Lett.*, **87**, 182902–04 (2005).
4. Srinivasan, G. et al. *Appl. Phys. A Mater. Sci. Process*, **78**, 721–728 (2004).
5. Bichurin, M. I., Petrov, V. M., and Srinivasan, G. *Phys. Rev. B*, 68, 054402–054414 (2003).
6. Srinivasan, G., Rasmussen, E. T., and Hayes, R. *Phys. Rev. B*, 67, 014418–014427 (2003).

7. Ramesh, R., Zavaliche, F., Chu, Y. H., Martin, L. W., Yang, S. Y., Cruz, M. P., and Barry, M. *Philos. Mag. Lett.*, **87**, 155–158 (2007).
8. Petrov, V. M., Srinivasan, G., Bichurin, M. I., and Gupta, A. *Phys. Rev. B*, 75, 224407–224416 (2007).
9. Spaldin, N. A. and Fiebig, M. *Science*, 309, 391–392 (2005).
10. Kimura, T., Goto, T., Shintani, H., Ishizaka, K., Arima, T., and Tokura, Y. *Nature*, **426**, 55–59 (London) (2003).
11. Lottermoser, T., Lonkai, T., Amann, U., Hohlwein, D., Ihringer, J., and Fiebig, M. *Nature*, **430**, 541–545 (2004).
12. Ramesh, R. and Spaldin, N. A. *Nature Materials*, **6**, 21–29 (2007).
13. Zhou, J. P., He, H. C., Shi, Z., Liu, G., and Nan, C. W. *J. Appl. Phys.*, **100**, 094106–094111 (2006).

8   Novel Inorganic-Organic Hybrid Resins as Biomaterials

P. P. Lizymol

CONTENTS

8.1   INTRODUCTION

The most popular oligomer currently used in dentistry for dental composite preparation based on a combination of a highly viscous bifunctional monomer Bisphenol A- Glycidyl methacrylate (Bis-GMA) mixed with low viscosity dimethacrylates like triethylene glycol dimethacrylate (TEGDMA). The resin is the binder for the filler and it is the continuous phase. From a clinical perspective, the dentist prefers visible light cure composites due to the convenience of visible light activation for polymerization. Dentistry has an ever expanding variety of restorative materials that require curing by photoinitiation. The most commonly used photoinitiator is (-) camphor quinone (CQ) which is a diketone with a λ_{max} at 473 nm along with a suitable amine as electron donor. The CQ has a yellowish color that makes harder to match the shades. The CQ forms a slow polymerization trajectory with only 50–75% initial conversion on light exposure [1] which improves slowly to reasonable levels. Also CQ needs an amine which functions as an electron donor. Presence of amine results to change in shade, loss of activity during storage and toxicity of the resultant restorative. Moreover, CQ has an unbleachable chromophore group and a limitation in its use remains discoloration of the set material. Exceeding a critical concentration limit, whereby unreacted molecules are not able to react with the co-initiator and return to the ground state results in yellow discoloration [2-5] which may result in an aesthetically undesirable restoration. In order to formulate bright white or translucent shades of resin based composites, some manufacturers use less CQ and/or alternative photoreactive species such as 2,4,6-trimethylbenzoyl-diphenylphosphine oxide (Lucirin-TPO) and 1-phenyl-1,2-propanedione (PPD), which has a maximum absorbance spectral peak in lower wavelength ranges (390 and 410 nm peak respectively) have been used in combination with CQ to lessen the yellowing effect and ultimately improve the curing efficiency of the polymerization system [6, 7] The high efficiency of the reaction coupled with the strong absorption and its stability toward hydrolysis has made TPO [8] as a commercial photoinitiator for curing polymer resins. The objective of the present study was to evaluate the photosensitization efficiency of TPO on a new organically modified ceramic resin and compare it with Bis-GMA in terms of properties such as depth of cure (DP), compressive strength (CS), diametral tensile strength (DTS), flexural strength (FS), flexural modulus (FM) and Vickers hardness number (VHN) of cured composite.

Synthesis, characterization, and applications of novel organically modified ceramic resins which are inorganic-organic hybrid materials are described here. These materials find application in electronic, optics, wear resistant coating, and tissue engineering and as dental restoratives. These materials showed excellent biocompatibility, surface hardness, and superior shrinkage characteristics in the oral cavity. Various aspects on the development of novel calcium containing resins and composites used as restorative material to replace damaged tooth which is mainly a natural composite of collagen and hydroxy apatite is presented. Effect of various photoinitiators on physical properties of photocured composites showed that photosensitization efficiency of TPO on dental resin polymerization of organically modified ceramic resin was rapid and more complete than the conventional photoinitiator, (-) CQ.

8.2 MATERIALS AND METHODS

8.2.1 Materials

Organically modified ceramic resin prepared by sol-gel process [9] of 3-methacryloxypropyltrimethoxysilane was used as the binder resin and purified silanated quartz was used as the filler for the preparation of restorative paste. Bis-GMA (Aldrich Chem. Co. Milwaukee, USA) was used as the control material. The TPO (Aldrich Chem. Co. Milwaukee, USA) was used as the photoinitiator. Other chemicals used were TEGDMA, 4-methoxy phenol, phenyl salicylate, 2-hydroxy-4 methoxy benzophenone and 2,6 di-tert-butyl-4-methyl phenol (all from Aldrich Chem. Co. Milwaukee, USA) which were acted as stabilizers, inhibitors, and so on.

8.2.2 Silanation of Filler

A coupling agent 3-methacryloxypropyltrimethoxysilane (Aldrich Chem. Co. Milwaukee, USA) was used directly without further purification. A 1% solution of 3-methacryloxypropyltrimethoxysilane in acetone with respect to filler was prepared and added to the filler. The mixture was stirred at 40°C in a water bath till the solvent evaporated completely. Then the filler was heated at 120°C for 1 hr in an air oven, cooled and used for preparing dental composite paste.

8.2.3 Dental Composite Paste Preparation

The TEGDMA was used as the diluent monomer for both organically modified ceramic and Bis-GMA resin. The Bis-GMA/TEGDMA ratio was 70/30 while organically modified ceramic resin/TEGDMA ratio was 50/50 due to its high viscosity. The TEGDMA content was adjusted to get equal flowability for both resin mixtures. To this mixture, 0.5% by weight of photoinitiator (TPO) inhibitor, and UV stabilizers (200 ppm-0.25%) were mixed. The prepared resin mixture was mixed with 280–300% of silanated quartz and 12% pyrogenic silica in a wear resistant agate mortar to obtain a uniform paste. The paste prepared using Bis-GMA was coded as BGQZ-TPO and organically modified ceramic resin was Ormo 48 TPO.

8.2.4 Evaluation of Dental Composites

Depth of Cure

Brass moulds with 3 mm diameter and 6 mm depth were used to prepare the samples for DP measurements. The mould was placed on a strip of transparent sheet on a glass slide and composite paste was packed in to the mould. A second strip of transparent sheet was placed on the top followed by a second glass slide. The mould and strip of film between the glass slides were pressed to displace excess material. The glass slide covering the upper strip was removed and the paste was exposed to visible light (Pro Lite, Caulk, Dentsply) for 60 s. The sample was then taken from the mould and the uncured part of the paste was removed using a metallic spatula. The depth of the cured part was measured accurately to the nearest millimeter using a digimatic vernier caliper of accuracy 0.01 mm and divided by two to get the DP. Mean and standard deviation were calculated.

Diametral Tensile Strength (DTS)

The DTS was determined using a reported procedure [9]. Cylindrical specimens, 6 mm diameter (d) × 3 mm height (L), were prepared by photo-curing using visible light cure unit (Pro Lite, Caulk, Dentsply). Intensity used was >300 mw/cm². Both sides of the samples were exposed to 60 s. The samples were stored in distilled water at 37 ± 1°C for 23 hr and at 22°C for 1 hr to attain room conditions prior to testing. Compressive force (P) was applied diametrically by a universal testing machine (Instron model 3365, High Wycombe, Buckinghamshire, UK) at a crosshead speed of 10.00 mm/min. The DTS in MPa was calculated using the following equation:

$$DTS = 2P/\pi DL$$

where P is the load in Newton's, D is the diameter and L is the thickness of the specimen in mm. Mean and standard deviation of six values were calculated.

Compressive Strength (CS)

Cylindrical specimens, 3 mm diameter × 6 mm height, were prepared using split type brass mold. Both sides of the samples were exposed to 60 s. The samples were stored in distilled water at $37 \pm 1°C$ for 23 hr and at 22°C for 1 hr to attain room conditions prior to testing. Compressive force (P) was applied along specimen height by a universal testing machine (Instron model 3365, High Wycombe, Buckinghamshire, UK) at a cross-head speed of 10.00 mm/min up to failure. The CS in MPa was calculated using the following equation:

$$CS = 4P/\pi D^2$$

where P is the load in Newtons, D is the diameter and L is the thickness of the specimen in mm. Mean and standard deviation of six values were calculated.

Flexural Strength (FS)

The FS was determined using a reported procedure [9]. Rectangular specimens, 25 mm length × 2 mm width (b) × 2 mm depth (h), were prepared using split type stainless steel mold. Specimens were cured with the light source for 180 s on each side. Five specimens for each resin composite were made. Compressive force (P) was applied along specimen length kept on a three-point transverse testing rig with span length (l = 20 mm) between the two supports by a testing machine (Instron model 3,365, High Wycombe, Buckinghamshire, UK), at a cross-head speed of 1.00 mm/min up to failure. The FS and FM in MPa were calculated using the following equation:

FS (MPa) = 3 FL/2 bd²

(FM) (MPa) = FL³/4 bd³D

where

F = load at break in Newtons.

L = length of the specimen between two metal rods at the base plate in mm.

b = width of the specimen in mm.

d = diameter of the specimen in mm.

D = deflection in mm at load F from computer data.

Mean value and standard deviation of five samples were calculated.

Vickers Hardness Number (VHN)

The samples used for VHN measurements were similar to those used for DTS measurements. Hardness was measured without polishing the surface of the specimen. The VHN was measured for each side of the sample using a Vickers micro hardness Tester (Model HMV 2, Shimadzu, Japan) using a reported procedure [7]. Vickers hardness was calculated from the following equation. The mean value of six measurements was taken as the VHN.

H_v = 0.1891 F/d²

where H_v = hardness number

F = Test load (N)

d = mean length of the indentation diagonal length (mm)

Effect of Exposure Time on VHN

Photo curable pastes were packed inside the cavity of 6 mm diameter and 3 mm height samples as those used for DTS samples. Then the upper side of the specimen was exposed to visible light (Pro Lite, Caulk, Dentsply) source (same one used for DTS and FS) for 10, 20, 30, 40, 50, and 60 s durations and hardness was measured on the light exposed surface. For all exposure duration, six measurements were taken. In order to find the effect of storage of cured samples on post polymerization, 40 s cured samples were stored at 37°C for 24 hr and 7 day and the hardness was measured as described.

Statistical Analysis

Statistical evaluation was done by means of one-way analysis of variance (ANOVA). The $P < 0.05$ was considered as significant.

8.3 RESULTS

The DP values of the composites are given in Table 1. Both BGQZ-TPO and Ormo 48 TPO have comparable DP values. The depth of polymerization for light activated rein based composites will govern the mechanical and biological properties and ultimately the longevity of the restoration [2]. The complications associated with ineffective curing include the increased risk of bulk and marginal fracture insufficient wear resistance poor color stability and decreased cell viability [2].

Table 1 shows that DTS values of BGQZ-TPO was (46.44 MPa) and Ormo 48 TPO was (36 MPa). Showed [7] that DTS of BGQZ with CQ initiator was 36.7 MPa and combination of two photoinitiators CQ and 1-phenyl-1,2-propane dione (PPD) was 44.2 MPa. The FS, FM, and VHN of BGQZ TPO are significantly higher (Table 1) than BGQZ CQ [7] DTS and FS of BGQZ TPO were found to be comparable with BGQZ PPD and BGQZ CQ + PPD [7].

TABLE 1 Properties of cured dental composites with TPO photoinitiator.

Property evaluated	Material used	
	BG QZ TPO	**Ormo48 TPO**
Depth of cure (mm)	1.617 ± 0.01	$1.56 + 0.005$
Compressive strength (MPa)	216 ± 26.5	281 ± 22
Diametral tensile strength (MPa)	46.4 ± 4.5	36.5 ± 1.32
Flexural strength (MPa)	116.2 ± 14	81.4 ± 13
Flexural modulus (MPa)	13353 ± 3515	10801 ± 4201
Vickers hardness number (Kg/mm^2)	59.5 ± 2.64	91.6 ± 2.85

The CS and VHN (Table 1) of Ormo 48 TPO are found to be higher than BG QZ TPO. All other properties such as FS, FM, DTS, and DP are higher for BGQZ TPO. The DTS, FS, FM, and VHN of Ormo 48 TPO (Table 1) and Ormo 48 CQ [9] were comparable with no statistically significant difference. Effect of exposure time on VHN of Ormo 48 TPO and BGQZ TPO were shown in Figure 1.

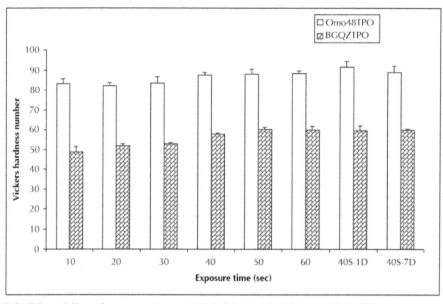

FIGURE 1   Effect of exposure time on VHN of Ormo 48 TPO and BGQZ TPO.

8.4   DISCUSSION

The TPO absorbs near uv (360–380nm) region and undergoes a photoinduced cleavage of the carbonyl–phosphinoyl bond from a triplet state. [10] Photolysis of TPO [8] affords 2,4,6-trimethyl benzoyl and diphenyl phosphine radicals via α-cleavage from a short lived triplet excited state. Studies [11] demonstrated that the mechanism of sensitization of TPO involves T–T energy transfer followed by formation of radicals by α-cleavage of the photoinitiators. In the case of organically modified ceramic based composite, Ormo 48 with both TPO (Table 1) and CQ (35.68 MPa) [9-12] initiators showed no significant difference in DTS, and VHN (91.6 ± 2.1) [9] at an exposure time of 60 s which indicated that the photosesitization efficiency of TPO is comparable with CQ for organically modified ceramic composite where as in the case of Bis-GMA based composites, the photo sensitizing efficiency of TPO is better than CQ [7] and comparable with PPD or combination of CQ and PPD as evidenced from the data in reference [7] and Table 1. Effect of exposure time on VHN of both Ormo 48 TPO and BGQZ TPO (Figure 1) showed that Ormo 48 TPO attained VHN of 83 at 10 s exposure while BGQZ TPO had 49. Both materials reached the maximum hardness at 40 s exposure. Further exposure and post curing had no effect on VHN. [9] Showed that both Bis-GMA (BGQZ CQ [9] and Ormo composites [9]) with CQ initiator, post curing at 37°C for 7 days after light exposure was required to reach the maximum value of VHN. Their hardness values at 10 s exposure time were 28 and 57 respectively [9]. Comparison of VHN at various exposure times of composites with both CQ

[9] and TPO showed that, monomer conversion was rapid for TPO and required lower exposure compared than CQ.

Hardness has been shown to be a good indicator of conversion of double bonds [13-15]. The greater the monomer conversion, the more will be the hardness. Therefore, hardness values can be used as an indirect measure of degree of conversion. However at longer exposure time (60 s) both TPO and (-)CQ showed similar photosensitization effect on Ormo 48 composite. But in the case of Bis-GMA based composite, at 60 s exposure time, TPO was found to be better than (-) CQ and comparable with (-) CQ + PPD combination.

8.5 CONCLUSION

The results of the study showed that photosensitization efficiency of TPO on dental resin polymerization of both Bis-GMA and organically modified ceramic resin was almost comparable. The present data and comparison with our published study showed that the photosensitization effect of TPO and CQ are comparable for organically modified ceramic composite at 1 min exposure time where as at lower exposure time of 10 s, composites with TPO showed better monomer conversion. However in the case of Bis-GMA based composites, TPO showed better efficiency than CQ and comparable with CQ + PPD combination. The TPO has the additional advantage that it can be used alone without the tertiary amine catalyst which was necessary for CQ and PPD.

KEYWORDS

- **Camphor quinone**
- **Depth of cure**
- **Diametral tensile strength**
- **Triethylene glycol dimethacrylate**
- **Vickers hardness number**

ACKNOWLEDGMENTS

The author is extremely grateful to The Director, Head, BMT Wing, and Scientist in Charge, Dental Products Laboratory, BMT Wing, Sree Chitra Tirunal Institute for Medical Sciences and Technology for providing facilities to carry out the project. Financial support from Department of Science and Technology, India under the scheme "Fast Track Proposals for Young Scientist" is gratefully acknowledged.

REFERENCES

1. Min-Huey Chena, Ci-Rong Chenc, Seng-Haw Hsuc, Shih-Po Sunc, and Wei-Fang Suc. Low shrinkage light curable nanocomposite for dental restorative material. *Dent Mater.*, **22**, 138–145 (2006).
2. Palin, W. M., Senyilmaz, D. P., Marquis, P. M., and Shortall, A. C. width potential for MOD resin composite molar restorations. *Dent. Mater.*, **24**, 1083–1094 (2008).
3. Ogunyinka, A., Palin, W. M., Shortall, A. C, and Marquis, P. M. Photoinitiation chemistry affects light transmission and degree of conversion of curing experimental dental resin composites. *Dent. Mater.*, **2**(3), 807–813 (2007).

4. Neumann, M., Miranda, W. J., Schmitt, C., Rueggeberg, F., and Correa, I. Molar extinction coefficients and the photon absorption efficiency of dental photoinitiators and light curing units. *J. Dent*, **33**, 525–32 (2005).
5. Nomoto, R. Effect of light wavelength on polymerization of light cured resins. *Dent Mater J.*, **16**, 60–73 (1997).
6. Park, Y., Chae, K., and Rawls, H. Development of a new photoinitiation system for dental light-cure composite resins. *Dent Mater*, **15**, 120–7 (1999).
7. Lizymol, P. P. and Kalliyanakrishnan, V. A. Comparative Study on the Efficiency of two photo initiators for polymerization of dental light cure composite resins *J. Appl, Polym. Sci.*, **107**(5), 3337–3342 (2008).
8. Sluggett, G. W., Turro, C., George, M. W., Koptyug, I. V., and Turro, N. J. (2,4,6-trimethylbenzoyl)diphenylphosphine Oxide Photochemistry. A Direct Time-Resolved Spectroscopic Study of Both Radical Fragments. *J. Am. Chem. Soc.*, **117**(18), 5148–5153 (1995).
9. Lizymol, P. P. Studies on New Organically Modified Ceramics Based Dental Restorative Resins. *J. Appl, Polym. Sci.*, **116**, 509–517 (2010).
10. Kolczak, U., Rist, G.,Dietliker, K., and Wirz, J. Reaction Mechanism of Monoacyl- and Bisacylphosphine Oxide Photoinitiators Studied by 31P-, 13C-, and 1H-CIDNP and ESR. *J. Am. Chem. Soc.*, **118**(27), 6477–6489 (1996).
11. Williams, R. M., Khudyakov, I. V., Purvis, M. B., Overton, B. J., and Turro, N. J. Direct and Sensitized Photolysis of Phosphine Oxide Polymerization Photoinitiators in the Presence and Absence of a Model Acrylate Monomer A Time Resolved EPR, Cure Monitor, and Photo DSC Study. *J. Phys. Chem. B*, **104**(44), 10437–10443 (2000).
12. Lizymol, P. P. Studies on shrinkage, depth of cure and cytotoxic behavior of novel organically modified ceramic based dental restorative resins. *J. Appl. Polym. Sci.*, **116**, 2645–2650 (2010).
13. David, J. R., Gomes, O. M., Gomes, J. C., Loguercio, A. D., and Reis, A. Effect of exposure time on curing efficiency of photopolymerizing unit equipped with light emitting diode. *J Oral Sci.*, **49**(1), 19–24 (2007).
14. Ferracane, J. L. Correlation between hardness and degree of conversion during setting reaction of unfilled dental restorative resin. *Dent Mater.*, **1**, 11–14 (1985).
15. Bouschlicher, M. R., Rueggeberg, F. A., and Wilson, B. M. Correlation of bottom to top surface microhardness and conversion ratios for a variety of resin composite compositions. *Oper. Dent.*, **29**, 698–704 (2004).

9 An Experimental *Investigation* on Hybrid Polymeric Composite Material Reinforced with Sisal and Acacia Fiber

K. Sakthivadivel and Dr. P. Govindasamy

CONTENTS

9.1   INTRODUCTION

In an advanced society like ours, we all depend on composite materials in some aspects of our lives. Composite materials have a long history of usage. Their beginnings are unknown, but all recorded history contains references to some form of composite material. For example, straw was used by Israelites to strengthen mud bricks. Plywood was used by ancient Egyptians when they realized that wood could be rearranged to achieve superior strength and resistant to thermal expansion as well as to swelling owing to the presence of moisture. Medieval swords and armor were constructed with layers of different materials.

Fiber glass developed in the late 1940s, [1] was the first modern composite and is still the most common. It makes up about 65% of all the composites produced today and is used for boat hulls, surfboards, sporting goods, swimming pool linings, building panels, and car bodies.

The demand for wood materials in construction of building field is on the rise. This leads to depletion of natural resources and encourages deforestation. The polymeric composite materials reinforced with natural fiber are being used as wood substitute and also for fabricating interior parts of automobiles. India is being endowed with an abundant availability of natural fibers, such as jute, sisal, pineapple, ramie, bamboo, banana, and so on has focused on the development of natural fiber composites. The development of natural fiber composites in India is based on a two pronged strategy of preventing depletion of forest resources as well as ensuring good economic returns for cultivation of natural fibers. In past investigation it has been noticed that the sisal, roselle, flax, and banana are very good in any one of the mechanical properties. We need a composite material which exhibits good mechanical properties such as compressive strength, tensile strength, impact strength, and so on. Sisal and roselle is good in tensile strength but moderate of compressive strength. For our study, we have taken the acacia fiber and polyester resin since it is good in compressive and tensile strength.

In this current investigation, the composite materials are reinforced with sisal and acacia fibers. These fibers were taken because of their good mechanical properties for our application purposes like helmet and car interior parts. Composite panels are fabricated by hand lay-up method. The mechanical properties such as tensile strength, compressive strength, and impact strength for different aspect ratios such as 0.5, 1.0, 2.0, 3.0, 4.0, and 5.0 cm are determined and compared with the mechanical properties of other natural fibers and so on.

9.2   SELECTION OF SISAL AND ACACIA FIBERS

This study deals with the polyester reinforced with sisal and acacia. The fibers are used as reinforcements and the unsaturated polyester resin is used as a matrix. These fibers are reinforcing agents because of their advantages over than the other fiber materials. The advantages of these fibers are easily available at a relatively low cost compared to other fibers and possess impressive mechanical properties. Sisal fiber is extracted from leaves of the sisal plant and the acacia fiber is new fiber that is extracted from the stem of a plant of the acacia, which is originally from Africa, but is also in Southern Asia. Before 19th centenary acacia fibers were used by farmers for lifting the heavy load because of their good mechanical properties, so we used this acacia fiber in composite materials.

9.2.1   Extraction of Sisal Fibers

Figure 1 shows the process of the extraction method of sisal fibers. The sisal leafs were cut from sisal plant and tied into bundles in bags. Then the bags containing the sisal leafs were retted in water for 5–8 days. The retted leafs were washed in running water and the top portion of the leaf was removed and then the fiber was cleaned and dried using the sunlight for 1 or 2 days.

(a)

(b)

(c)

FIGURE 1   Extraction method of sisal fibers (a) Naturally growing sisal plant, (b) After tied the sisal leafs in bags, retting in water for 5–8 days, and (c) Dried in sunlight.

9.2.2 Extraction of Acacia Fibers

Figure 2 shows the extraction method of acacia fibers. Cut the acacia plants and the stalk were tied into bundles and retted in water for 8–15 days. The retted stem of the acacia stem skin was washed in running water. Then the fibers were removed from stem skin, cleaned, and dried using the sunlight.

(a)

(b)

(c)

(d)

FIGURE 2   Extraction methods of acacia fibers (a) Naturally growing acacia plant, (b) After tied the acacia stems skin in bags, retting in water for 8–15 days, (c) Dried in sunlight, and (d) Finally got the acacia fibers.

9.3   SELECTION OF RESIN (POLYESTER)

Polymers generally act as a good binder for fibers. Their availability coupled with their lower cost has provoked the selection of polymer as the binder for these fibers [2]. Unsaturated polyester offers the advantage of easy moldability, better handling, and flow properties. Easy fabrication and better mixing of polyester provoke their usage. They have a low density of 1,150 kg/m^3 adding to our main objective of fabricating a low weight composite.

9.4   EXPERIMENTAL PROCEDURES

The experimental procedure deals with the fabrication and testing of composite materials. The fabrication process consists of fabricating the composite by using hand layup method. The testing process consists of mechanical property testing [3].

9.4.1   Hand Lay-up Method

The hand lay-up technique is the simplest and most commonly used method for the manufacture of both small and large reinforced products. It is used in applications where production volume is low and other forms of production would be prohibitive because of costs or size of equipments. The technique also called contact lay-up is an open mold method of molding thermosetting resins (polyesters and epoxies) in association with fibers.

Initially the percentages of resin, fibers, accelerator, and catalyst are determined for optimum weight percentages so that strength of the composite is notable and worthy. Here we have found that those percentages are 60% resin, 2% accelerator, 8% catalyst, and 30% of sisal and acacia fibers in equal proportions. First the chopped fiber as per the aspect ratio is laid over the acrylic sheet where an ASTM rubber of 10 mm thickness is placed over the sheet cut down to desired dimensions. After laying of fibers are over, another sheet is placed over the rubber and the sides are sealed to protect any leakages using tape then clips are used to hold the sheets together at a definite pressure. Once these things are over the resin mixed with the said percentage of components has to be poured between the space of two plates and after a curing time of about 5–7 hr the composite can be obtained. The hand lay-up method is as shown in Figure 3.

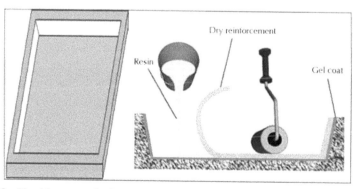

FIGURE 3   Hand lay-up method.

9.4.2 Mechanical Property Testing

The following mechanical properties of the hybrid polymer composite materials reinforced with sisal and acacia for different aspect ratio of 0.5, 1.0, 2.0, 3.0, 4.0, and 5.0 cm were determined during this investigation [4].

   (1) Tensile strength
   (2) Compressive strength
   (3) Impact strength

Tensile Test

Tensile test was carried out by using electronic tensometer, model photo view of PC 2000 and testing speed is 5 mm/min. The Figure 4 shows the electronic tensometer. The specimen size is 165 × 9 × 5 mm as per ASTM D3039/D3039M, ISO 527-4 (test method for tensile properties of polymer matrix).The Figure 5 shows the specimen size of the composite material.

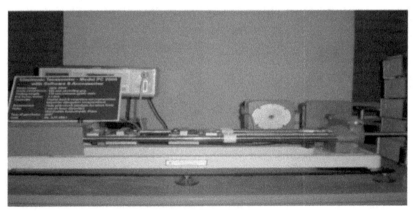

FIGURE 4   Photo view of electronic tensometer.

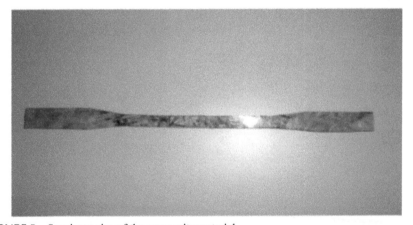

FIGURE 5   Specimen size of the composite material.

Electronic tensometer specification:
    Force range: upto 20 kN
    Cross Head Travel: 525 mm excluding grip
    Testing length: 135 mm between guide rods
    D.C Servo motor: 1.1 Nm
    Controller: Load and extension microprocess

Compression Test

Compression test was carried out using universal testing machine. The specimen size is 80 × 10 × 10 mm as per the ASTM D 695.

    Universal testing machine specification:

- PC based control system
- Max. load of 40 kN
- Graphical outputs.

Impact Test

Impact test is done by applying a sudden impact load over the composite. Impact test was carried out by using izod and charpy impact test machine. The specimen size is 75 × 10 × 10 mm and 35 × 10 × 10 mm respectively as per the ASTM D4812, ISO 180 and 179.

9.5   DISCUSSION AND RESULTS

This study deals with the testing results of composite materials. The various tests performed in mechanical testing properties are:

(1) Tensile test
(2) Compression test
(3) Impact test

9.5.1   Tensile Test

Tensile test is done by applying tensile load on the body by using the electronic tensometer. The test is contacted for the hybrid polymer composite material reinforced with sisal and acacia fibers for different aspect ratio of 0.5, 1.0, 2.0, 3.0, 4.0, and 5.0 cm.

    The various readings taken are:

(1) Peak load
(2) Breaking load
(3) Maximum displacement
(4) Tensile strength
(5) Percentage of elongation

The tensile test result is shown in Table 1 indicates that when the aspect ratio of fibers increases the experimental parameters such as peak load, breaking load, maximum displacement, tensile strength, and percentage of elongation are increasing.

TABLE 1   Tensile test results of hybrid polymer composite material reinforced with sisal and acacia fiber having different aspect ratio.

Length of Fibers (aspect ratio)	Peak Load N	Breaking Load N	Max Displacement mm	Tensile Strength N/mm²	% of Elongation
G. P. Resin	722.19	500.16	2.07	16.04	3.568
0.5 cm	1137.60	790.80	2.9	25.28	5.000
1.0 cm	1260.22	856.22	3.56	28.00	6.137
2.0 cm	1398.36	941.13	4.39	31.07	7.568
3.0 cm	1462.25	1028.39	5.02	32.49	8.655
4.0 cm	1542.17	1102.37	5.81	34.27	10.017
5.0 cm	1564.24	1118.94	5.88	34.76	10.13

The graph plotted in Figure 6 represents the tensile strength *versus* composite designation of different aspect ratio of fibers obtained after conducting the tensile test on hybrid polymer composite material reinforced with sisal and acacia. The graph shows when the aspect ratio of fibers increases the tensile strength is increase. The critical strength of tensile is 34.76 N/mm² of aspect ratio 5.0 cm.

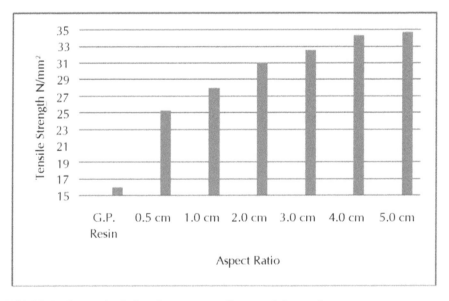

FIGURE 6   Composite designation *versus* tensile strength for tensile test.

9.5.2 Compression Test

Compression test is done by applying compression load on the body by using the universal testing machine. The test is contacted for the hybrid polymer composite material reinforced with sisal and acacia for different aspect ratio of 0.5, 1.0, 2.0, 3.0, 4.0, and 5.0 cm.

The various readings taken are:

(1) Peak load

(2) Breaking load

(3) Maximum displacement

(4) Compressive strength

The compression test result is shown in Table 2 indicates that when the aspect ratio of fibers increases the experimental parameters such as peak load, breaking load, compressive strength are increasing and maximum displacement is decreases.

TABLE 2   Compression test results of hybrid polymer composite material reinforced with sisal and acacia fiber having different aspect ratio.

Length of Fibers (aspect ratio)	Peak Load N	Breaking Load N	Max Displacement mm	Compressive Strength N/mm²
G. P. Resin	2501.55	1962	1.97	25.015
0.5 cm	5298.47	3539.05	0.81	52.984
1.0 cm	5909.34	4358.48	0.67	59.093
2.0 cm	6855.42	4795.91	0.39	68.554
3.0 cm	7775.60	5124.64	0.21	77.756
4.0 cm	8045.96	5377.94	0.18	80.459
5.0 cm	8112.08	5421.104	0.184	81.12

The graph plotted in Figure 7 represents the composite designation *versus* compressive strength of different aspect ratio of fibers obtained after conducting the compression test on hybrid polymer composite materials reinforced with sisal and acacia. The graph shows when the aspect ratio of fibers increases the compressive strength is increase [5]. The critical strength of compressive is 81.12 N/mm² of aspect ratio 5.0 cm.

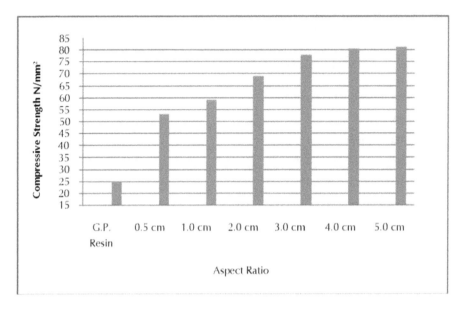

FIGURE 7 Composite designation *versus* compressive strength for compression test.

9.5.3 Impact Test

Impact test is done by applying a sudden impact load over the composite. Impact test was carried out by using izod and charpy impact test machine. The test is contacted for the hybrid polymer composite material reinforced with sisal and acacia for different aspect ratio of 0.5, 1.0, 2.0, 3.0, 4.0, and 5.0 cm.

The impact test result shown in Table 3 indicates that when the aspect ratio of fibers increases the experimental parameters such as impact strength of charpy and izod are increasing.

TABLE 3 Impact test results of hybrid polymer composite material reinforced with sisal and acacia fiber having different aspect ratio.

Length of Fibers (aspect ratio)	Impact Strength J/Sq.cm(Charpy)	Impact Strength J/Sq.cm(Izod)
G. P. Resin	8.20	9.0
0.5 cm	11.00	12.50
1.0 cm	11.90	13.20
2.0 cm	12.80	13.70
3.0 cm	13.90	14.90
4.0 cm	15.00	16.00
5.0 cm	15.10	16.20

The graph plotted in Figure 8 represents the impact strength *versus* composite designation of different aspect ratio of fibers obtained after conducting the impact test on hybrid polymer composite materials reinforced with sisal and acacia. The graph shows when the aspect ratio of fibers increases the impact strength is increase. The critical strength of impact charpy is 15.10 J/cm² and impact izod is 16.20 J/cm² of aspect ratio 5.0 cm.

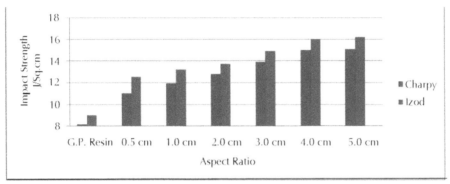

FIGURE 8   Impact strength *versus* composite designation for impact test.

9.6   CONCLUSION

Characteristic and mechanical properties of the natural fibers like banana, flax, jute, and so on were studied. This shows good in any one of the mechanical properties on nature [6]. We have taken sisal and acacia fibers to study its characteristics and mechanical strength. It is found that acacia fiber has outstanding mechanical properties like tensile, impact, and compressive strength comparing those of the other fibers. Hybrid composites were fabricated using polyester resin and reinforced with sisal and roselle fibers. The following results were obtained follows:

(1) Fabricated the hybrid polymer composite materials reinforced with sisal and acacia fibers with varying aspect ratio of fibers by using hand lay-up method.

(2) The mechanical properties such as tensile strength, flexural strength, compressive strength, and impact strength for different aspect ratio of fibers are evaluated and compared the mechanical properties with other natural fibers.

(3) The tensile test result shows that, the aspect ratio of 5.0 cm provides highest tensile strength 34.76 N/mm².

(4) The compression test result shows that the aspect ratio of 5.0 cm provides highest compressive strength 81.12 N/mm².

(5) The impact test result shows that the impact strength increases as the aspect ratio increases, so the aspect ratio of 5.0 cm provides highest impact strength of charpy is 15.10 J/cm² and izod is 16.20 J/cm².

(6) The tensile, compression, and impact test result shows that the aspect ratio of 5.0 cm provides highest mechanical properties compared to the aspect ratio of 0.5, 1.0, 2.0, 3.0, and 4.0 cm.

(7) The investigation of result shows that the sisal and acacia fibers are have good strength in all of the mechanical properties like tensile, compressive, and impact strength.

(8) In this investigation, while increasing the length of the fibers it gives the high strength in all the mechanical properties.

(9) Mechanical advantage of this fiber is best alternative to most of wood and plastics materials.

KEYWORDS

- Acacia
- Hand lay-up process
- Mechanical properties
- Polyester resin
- Sisal

REFERENCES

1. Powell R. M. Utilization of Natural Fibers in Plastic Composites. *Lingocellulosic Plastic Composites*, **4**(27), (1994).

2. Sastra, H. Y. Flexural Properties of Agenta Fiber Reinforced Composites. *American Journal of Applied Sciences*, (Special Issue), 21–24 (2005).

3. Harish, S., Michael, P. D., Bensely, A., and Rajadurai, A. Mechanical property evaluation of natural fiber coir composite. *Material Characterization*, **60**, 44–49 (2009).

4. Sakthivadivel, K. and Sathiamurthi, P. Fabrication and Analysis of Natural Fiber Composite Materials (Sisal and Roselle). *ICRAME*, ISBN: 978-81-907917-3-1, **2**, 328–335 (2010).

5. Antich, P. and Vazquez, A. Mechanical behavior of high impact polystyrene reinforced with short sisal fibers. *Applied science and manufacturing*, **37**, 139–150 (2006).

6. Thi-Thu-Loan Doan and Shang-Lin Gao. Jute/polypropylene composites effect of matrix modification. *Composites Science and Technology*, **66**, 952–963 (2006).

10 Modernism of Ni–WC Nanocomposite Coatings Prepared from Pulse and Direct Current Methods

B. Ranjith and G. Paruthimal Kalaignan

CONTENTS

10.1 INTRODUCTION

Composite electrodeposition technology is one of the methods used to fabricate metal matrix composite coating materials. The basic process is conducted in the composite electrodepositing solution in which a large amount of insoluble micropowders are suspended by stirring the solution and the micropowders are embedded into the coating during the direct current (DC) and pulse current (PC) electrodeposition process of metal matrix. The excellent performance of electrodeposited composite coatings, especially in mechanical and chemical processes [1, 2], account for the widespread use of the process. It is well known that nickel electro composite coatings can be easily obtained from a nickel bath and number of studies on electro composites by using a nickel bath have been reported from a technological point of view [3-12]. Pulse electrodeposition has proven to be one of the most effective methods in fabrication of metal coatings. As compared with traditional DC electrodeposition, pulse electrodeposition offers more process controllable parameters, which can be adjusted independently and can withstand much higher instantaneous current densities. Therefore, metal coatings fabricated by pulse electrodeposition possess more unique compositions and microstructures when compared to DC electrodeposition. Electrodeposition of composite coatings, based on hard particles dispersed in a metallic matrix, is gaining importance for potential engineering applications [13]. The second phase can be hard oxide (Al_2O_3, TiO_2, SiO_2), carbides particles (SiC, WC), diamond, solid lubricate (PTFE, graphite, MoS_2), and even liquid containing microcapsules [14]. A well known application of codeposition to improve the corrosion resistance of coatings is the production of microporous chromium layers by codepositon of nonconducting particles on the underlying Ni layer [15]. Corrosion of Ni occurs over an increased surface resulting in a smaller depth of attack. In an oxygen-enriched atmosphere, Ni–Al_2O_3 composite coatings possess better oxidation resistance than unreinforced nickel [16]. Improvement in the wear resistance has also been reported for electrodeposited Ni–SiC composites [17]. With these considerations, many investigators have successfully codeposited hard particles (like Al_2O_3, TiO_2, SiC, WC, Cr_3C_2, TiC, diamond, etc.) in a range of metal matrices such as Ni, Cr, Co, Re, and so on [18]. Tungsten carbide (WC) or WC–Co is a technologically important material and has been widely used as cutting tools, rock drills, punches, and wear resistant coating materials [19]. For example, WC–Co coatings have been extensively used in industry to prevent wear, erosion, and corrosion of many metallic components. Although extremely important, engineering WC thin films on various important materials remains an open area in nanotechnology.

In this chapter, Ni–WC composite coatings were prepared on mild steel substrate by DC and PC methods. Their microstructure and properties were also investigated by various techniques. The results showed that PC could improve considerably the performances of Ni–WC composite coatings.

The Ni–WC nanocomposite coatings were prepared under DC and PC using acetate bath. The microstructure and corrosion resistance of the coatings were characterized by means of XRD, SEM, AFM, and EIS. The results showed that the microstructure and performances of the coatings were greatly affected by WC content on the deposits prepared by DC and PC methods. The microhardness and corrosion resistance were enhanced in the optimum percentage of WC composite coatings. The PC composite coatings were exhibited compact surface, higher microhardness, and good corrosion resistance compared with that of the DC composite coatings.

10.2 EXPERIMENTAL

10.2.1 Electrode Pretreatment and Electrolyte Preparation

Mild steel plate has been used as a substrate for depositing pure Ni and Ni–WC composite coatings. The substrates were subjected to pretreatment to remove organic and inorganic impurities. The MS plate cathode size of $2.5 \times 2.5 \times 0.2$ cm was dipped in 10% HCl for 10 min and then cleaned with distilled water followed by drying and degreasing with trichloroethylene. Pure Nickel bar (99.9%) of size $5 \times 5 \times 0.5$ cm was used as anode.

10.2.2 Bath Preparation and Optimization

The plating bath composition, current density and pH were optimized with the help of throwing power and Hull cell methods. Bath solutions were prepared by using AnalaR grade chemicals and triple distilled water. The optimized bath composition and electroplating conditions for DC and PC methods are given in Table 1.

TABLE 1 The optimized bath composition and electroplating conditions.

Electrolyte Composition		Electroplating Parameters	
		Direct, Pulse and	
		Pulse Reverse current	
Nickel Acetate	: 150 g/l	Current density	: 0.4 A/inch ^{2}
Nickel Chloride	: 45 g/l	pH	: 4.5
Cobalt acetate	: 15 g/l	Temperature	: 30 ^{0} C
Boric acid	: 35 g /l	Plating Time	: 90 min
Tungesten Carbide	: 2g – 8g/l	Duty Cycle	: 30 cycle
–		Stirring Speed	: 150 r/min

10.2.3 Direct Current and Pulse Current Plating

The pretreated polished and degreased MS plate cathode of size 2.5×2.5 cm was dipped electrolyte bath at 25°C. Nickel bar of size 5×5 cm act as anode. The various composition of WC was added (2–8 g/l) into the bath. After the addition of WC, the bath was stirred well before and during the plating time of 30 min using laboratory stirrer. The DC and PC experiment were carried out using pulse rectifier (Komal agencies, Mumbai).

10.2.4 Microhardness, SEM, XRD, and AFM

Microhardness was measured using MH 6 everyone hardness tester (Hong Kong). The hardness values were measured in three different locations for each sample. The SEM (HITACHI S-570, Japan) determined surface morphology of the composite coatings. Phase structure of the coatings was analyzed by the XRD (X' Pert PRO diffractometer with Cu Kα radiation). The samples were scanned between 20 and 100° (2θ) at a scan rate of 1°/min. The surface morphologies of the target composite coatings were observed using an atomic force microscope (AFM). The samples were scanned by using dicp – II veeco.

10.2.5   Electrochemical Measurements

The corrosion behavior of pure Nickel and the Ni–WC composite coatings was characterized by potentiodynamic polarization and AC impedance measurements.

Potentiodynamic Polarization Studies

The electrochemical measurements were carried out in a conventional three electrode cell at 25°C. The MS specimen was masked with lacquer to expose 1 cm² area which served as the working electrode. Pt foil (1 cm²) was used as the counter electrode and a saturated calomel electrode (SCE) as the reference. The EG and G—Auto lab Analyzer (Model: 6310) was employed for the polarization studies and the potential of the working electrode was varied with respect to SCE. The mild steel specimens were immersed in the test solution of 3.5% Na–Cl and allowed to attain a steady potential value. The potentiodynamic polarization was carried out from 0.75 to 1.25 V with respect to the OCP at a scan rate of 2 mV/s. The potential E (V *versus* SCE) was plotted against log I (A cm^{-2}) to obtain polarization curve. From this polarization curves, the corrosion potential (E_{corr}) and corrosion current (i_{corr}) of the specimens were obtained using the Tafel extrapolation method.

Electrochemical Impedance Spectrum

The electrochemical impedance spectrum for electrolytes with various amount of SiC particles were measured using an EG and G—Auto lab Analyzer (Model: 6310). The same three electrode cell setup was used for this experiment. The EIS was carried out between 10 and 0.01 Hz frequency ranges (rms amplitude – 5 mV).

10.3   DISCUSSION AND RESULTS

10.3.1   Effect of Particle Concentration on Codeposition

The incorporation of WC in the Nickel deposit with different concentrations (2–8 g/l) of WC particles suspended in the bath at current density 6.4 A/dm², 30°C is shown as Figure 1. The wt% of WC in the deposit increases sharply with increasing the concentration of WC in the bath and attains the maximum values at 6 g/l WC suspension in the bath at 6.4 A/dm² respectively. With the further additions, the wt% of WC slightly decreases. There is measurable significance change in the incorporation of WC particle in the Nickel matrix during the deposition of DC and PC. The curve is quite similar to the well known Langmuir adsorption isotherms, supporting a mechanism based on an adsorption effect. The codeposition of WC nanoparticles on the cathode surface was suggested by two step adsorption model [20]. Once the particles are adsorbed, metals begin to building around the cathode slowly, encapsulating and incorporating the particles.

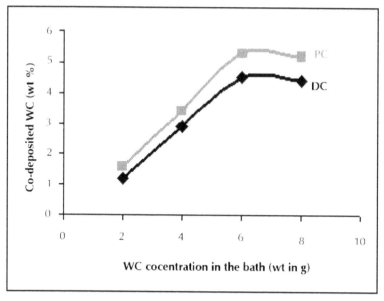

FIGURE 1 Effect of WC concentrations in the bath on wt% of WC in the nanocomposite coatings prepared by DC and PC.

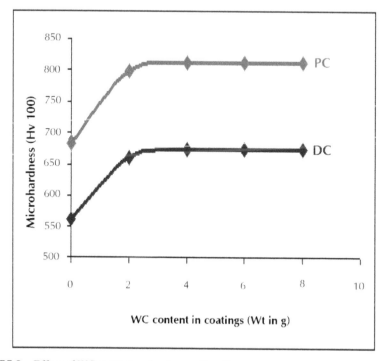

FIGURE 2 Effect of WC content on hardness of coatings.

10.3.2   Microhardness

Figure 2 represents the Vickers hardness *versus* WC content in coatings (wt in g) for the Ni–WC nanocomposites. This figure illustrate the influence of WC content on the hardness of the composite coatings prepared in the same solution by DC and PC methods at the plating time of 30 min respectively. It can be seen that, the hardness of coatings was improved and decrease the macro residual stress of the coatings through the introduction of WC particles. Figure 1 also indicates that increasing the concentrations of WC in the bath increased the coating hardness, which implies that, more WC was incorporated into the nickel matrix. All the Ni–WC nanocomposite coatings have higher micro hardness, compared to pure Ni coating. The hardness of the PC composite coating is higher than that of DC composite coatings. Accordingly, the higher microhardness value of the Ni–WC nanocomposites may be due to the decrease of the grain size of Nickel matrix of the composites, which is favored by the WC particles. With the grain refinement of Nickel matrix, the load carrying ability and the resistance for plastic deformation [21-24] increase. It is also known that the hardness and other mechanical properties of metal matrix composites depend in general on the amount and size of the dispersed phase, apart from the mechanical characteristics of the matrix.

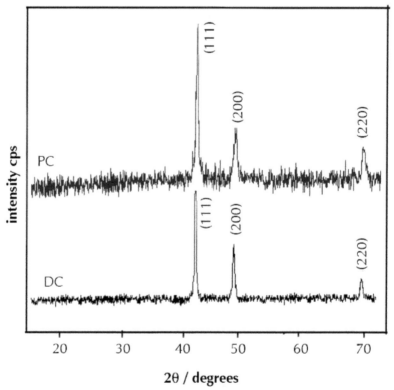

FIGURE 3   Phase structure of Ni–WC nanocomposite coatings.

10 3.3   Structural Analysis by XRD Measurements

The XRD patterns of the Ni–WC coatings were illustrated in Figure 3. The DC and PC nanocomposite coatings showed merely the fcc structure of the Ni matrix. Compared with the DC coatings, the widths of diffraction peaks (1 1 1), (2 0 0), and (2 2 0) of PC coatings are all significantly wider than those of DC coatings, indicating the fine grained texture of PC coatings. The results of the present investigation showed that, by means of PC technique, it could achieve fine grain, smooth surface and compact composite coatings. The results may be attributed to increasing electrochemical polarization of cathode during the on time of PC period, which decreases the nucleation energy of the metal deposition on the electrode surface and increases nucleation rate. As a consequence of the high nucleation, it leads to increase the number of nucleation centre [25]. Consequently, the texture of PC coatings possesses fine grain and compact microstructure.

10.3.4   Surface Morphology by SEM Studies

The SEM micrographs of pure nickel and Ni–WC nanocomposites are shown in Figure 4 (a, b, c). The pure nickel films from the acetate bath showed a well defined field-oriented texture (FT) type structure (Figure 4(a)). Due to the addition of WC, particles to the bath the microstructure of the nickel matrix changed from columnar to granular (Figure 4(c)). In PC method, Ni–WC nanocomposites exhibited a granular and fine uniform structure.

(a)

FIGURE 4   *(Continued)*

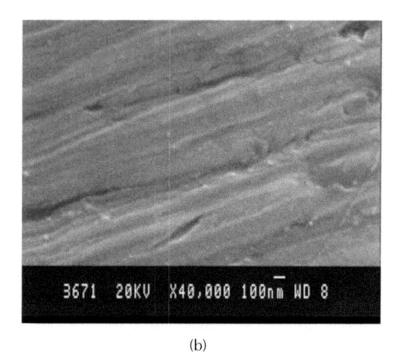

(b)

(c)

FIGURE 4   SEM photographs of (a) Pure Ni (b) Ni–WC (6 g/l) (c) Ni–WC (6 g/l) nanocomposite coatings prepared by DC and PC methods.

10.3.5 AFM Investigation of the Surface Morphology

In order to investigate the microstructure of Ni–WC nanocomposite coating more detailed, scanning analysis and measurements were conducted using AFM method (see Figure 5 (a) and (b)). The microstructure of Ni–WC (6 g/l) appears to be influenced by direct and pulse composite coatings. The microstructure of Ni–WC (6 g/l) nanocomposite coating prepared in PC method has a more uniform and fine structure than that of Ni–WC (6 g/l) nanocomposite coating prepared in DC method.

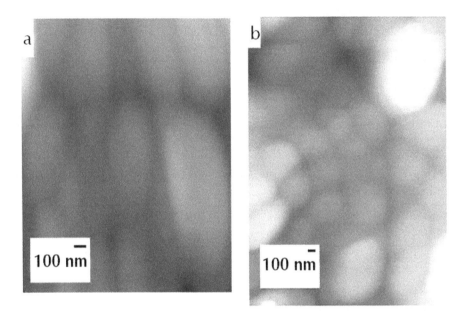

FIGURE 5 AFM Photographs of Ni–WC (6 g/l) nanocomposite coatings prepared by (a) DC (b) PC methods.

10.3.6 Electrochemical Measurements

Electrochemical Impedance Spectroscopy

The corrosion resistance property of the electrodeposited pure Ni and Ni–WC nanocomposite coatings were measured by electrochemical impedance spectroscopy using a three electrode cell assembly. Impedance spectra were recorded for pure Ni and Ni–WC (2 g/l, 4 g/l, and 6 g/l) composite coated samples prepared in DC and PC method. From the Figure 6 (a) and (b), the Nyquist plots value of charge transfer resistance (R_{ct}) and double layer capacitance (C_{dl}) were calculated. The charge transfer resistance (R_{ct}) values for Ni–WC nanocomposite coatings increased and the double layer capacitance values decreased with WC in the composite coatings Table 2. The analysis confirmed that the incorporation of WC enhances the corrosion resistance of composite coated samples. It revealed that Ni–WC nanocomposite coating from PC is more corrosion resistant than Ni–WC nanocomposite coating from DC method.

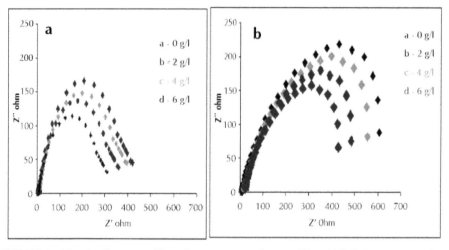

FIGURE 6 Nyquist diagrams of impedance spectrum for pure Ni and Ni–WC nanocomposite coatings prepared by (a) DC (b) PC methods.

TABLE 2 Parameters derived from EIS for Pure Ni and Ni–WC nanocomposite coatings prepared by DC and PC method.

Methods	Samples	R_{ct} (ohms)	C_{dl} (FCm^{-2}) ‰ 10^{-5}
Direct Current (DC)			
	Pure Ni	348	4.94
	Ni–WC (2 g/l)	371	4.59
	Ni–WC (4 g/l)	381	4.24
	Ni–WC (6 g/l)	415	3.99
Pulse Current (PC)			
	Pure Ni	494	4.07
	Ni–WC (2 g/l)	510	4.01
	Ni–WC (4 g/l)	596	3. 87
	Ni–WC (6 g/l)	684	3.10

Tafel Polarization Behavior of Ni–WC Nanocomposite Coating

The corrosion potentials (E_{corr}), the corrosion current, (i_{corr}), corrosion rate for electrodeposited pure Ni and Ni–WC nanocomposite coatings were calculated from the Tafel polarization curves Figure 7 (a) and (b). The corrosion current (i_{corr}) decreased in all the Ni–WC nanocomposite coatings compared to electrodeposite pure Ni coating Table 3. The corrosion rate of electrodeposited pure Ni and Ni–WC nanocomposite coatings was also calculated and the dependence of corrosion rate on the wt% of WC particles in the Ni–WC nanocomposite coating. It revealed that Ni–WC nanocomposite coating prepared from PC method was more corrosion resistant than Ni–WC nanocomposite coating prepared from DC method.

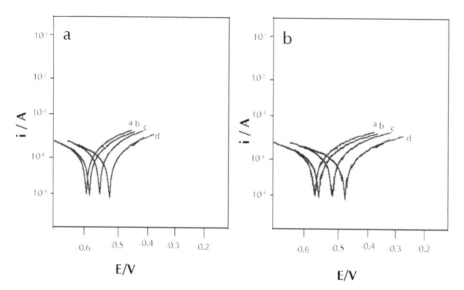

FIGURE 7 Potentiodynamic polarization curves for pure Ni and Ni–WC nanocomposite coatings prepared by (a) DC (b) PC methods.

TABLE 3 Parameters derived from Polarization curves for Pure Ni and Ni–WC nanocomposite coatings prepared by DC and PC method.

Methods	Samples	E_{corr} (V)	I_{corr} $(10^{-6}A)$
Direct Current (DC)	Pure Ni	− 0.683	5.804
	Ni–WC (2 g/l)	−0.526	4.901
	Ni–WC (4 g/l)	−0.519	4.781
	Ni–WC (6g/l)	−0.483	4.325

TABLE 3   *(Continued)*

Methods	Samples	E_{corr} (V)	I_{corr} $(10^{-6}A)$
Pulse Current (PC)	Pure Ni	−0.552	5.309
	Ni–WC (2 g/l)	−0.483	4.380
	Ni–WC (4 g/l)	−0.459	4.331
	Ni–WC (6g/l)	−0.422	4.265

10.4   CONCLUSION

(1) The PC could obtain Ni–WC composite coatings with finer crystal grain, smoother surface and more homogeneous microstructure. Under the experimental conditions, WC particles uniformly distributed throughout the PC composite coatings and combined very well with the Nickel matrix.

(2) By application of PC, the hardness and corrosion resistance were superior to DC coatings, so the PC technique is a promising method for modification and improvement of surface.

KEYWORDS

- **Corrosion current**
- **Electrodeposition Techniques**
- **Microhardness**
- **Nanocomposite coating**
- **Saturated calomel electrode**

REFERENCE

1. Ferkel, H., Muller, B., and Riehemanm, W. *Mater. Sci. Eng. A.*, **474**, 234–236 (1997).
2. Benea, L., Bonora, P. L., Borello, A., and Martelli, S. *J. Electrochem. Soc.*, **148**, 461 (2001).
3. Oberle, R., Scanlon, M. R., Cammarata, R. C., and Searson, P. C. *Appl. Phys. Lett.*, **66**, 19 (1995).
4. Peipmann, R., Thomas, J., and Bund, A. *Electrochim. Acta.*, **52**, 5808 (2007).
5. Shao, I., Vereecken, P. M., Chien, C. L., Searson, P. C., and Cammarata, R. C. *J. Mater. Res.*, **17**, 1412 (2002).
6. Tu, W. Y., Xu, B. S., and Dong, S. Y. *Mater. Lett.*, **60**, 1247 (2006).
7. Vaezi, M. R., Sadrnezhaad, S. K., and Nikzad, L. *Colloids Surf. A.*, **315**, 176 (2008).
8. Wei, X., Dong, H., Lee, C. H., and Jiang, K. *Mater. Lett.*, **62**, 1916 (2008).
9. Zhou, Y., Peng, X., and Wang, F. *Scr. Mater.*, **55**, 1039 (2006).
10. Zhou, Y., Zhang, H., and Qian, B. *Appl. Surf. Sci.*, **253**, 8335 (2007).

11. Zhou, Y. and Zhang, H. *Cailiao Rechuli Xuebao/Transactions of Materials and Heat Treatment*, **29**, 154 (2008).
12. Aruna, S. T. and Rajam, K. S. *Scr. Mater.*, **48**, 507 (2003).
13. Ghorbani, M., Mazaheri, M., Khangholi, K., and Kharazi, Y. *Surf. Coat. Technol.*, **148**, 71 (2001).
14. Alexandridou, S., Kiparissides, C., Fransaer, J., and Celis, J. P. *Surf. Coat. Technol.*, **71**(3), 267 (1995).
15. Narayan, R. and Chttopadyay, S. *Surf. Technol.*, **16**, 227 (1982).
16. Stott, F. A. and Ashby, D. *J. Corros. Sci.*, **18**, 183 (1978).
17. Hou, K. H., Ger, M. D., Wang, L. M., and Ke, S. T. *Wear*, **253**(9–10), 994 (2002).
18. Nwoko, V. O. and Shreir, L. L. *J. Appl. Electrochem.*, **3**(2), 137 (1973).
19. Ban, Z. G. and Shaw, L. L. *J. Mater. Sci.*, **37**, 3397 (2002).
20. Guglielmi, N. *J. Electrochem. Soc.*, **119**, 1009 (1972).
21. Benea, L., Bonora, P. L., and Borello, A. et al. *Wear*, **249**, 995–1003 (2002).
22. Lekka, M., Kouloumbi, N., Gajo, M., and Bonora, P. L. *Electrochim. Acta*, **50**, 4551–4556 (2005).
23. Srivastava, M., William Grips, V. K., and Rajam, K. S. *Appl. Surf. Sci.*, **253**, 3814–3824 (2007).
24. Li-Ping, W., Yan, G., Tao, X., and Quan Ji, X. *Mater. Chem. Phys.*, **99**, 96–103 (2006).
25. Zhang, F., Jing, T. F., and Qiao, G. Y. *Plat. Finish*, **4**, 1 (in Chinese) (2001).

11 Effect of Microfiller Addition on Physico-mechanical Properties of Glass Fabric Reinforced Epoxy Composite Systems

B. Suresha, Y. S. Varadarajan, and G. Ravichandran

CONTENTS

11.1 INTRODUCTION

Many of our modern technologies require materials with unusual combination of properties that cannot be met by the conventional metal alloys, ceramics, and polymeric materials. This is especially true for materials that are needed for aerospace, military, automotive, and transport applications. For example, aircraft engineers are increasingly searching for structural materials that have low densities, are strong, and stiff and abrasion and impact resistant, and are not easily corroded.

Material property combinations and ranges have been, and are yet being, extended by the development of composite materials. Generally speaking, a composite is considered to be any multiphase material that exhibits a significant proportion of the properties of both constituent phases such that a better combination of properties is realized. According to this principle of combined action, better property combinations are designed by the judicious combination of two or more distinct materials.

A composite is a multiphase material that is artificially made, as opposed to one that occurs or forms naturally. Composite materials provide an opportunity to combine different properties and design material systems for applications requiring multiple functionalities. Many composite materials are composed of two phases: one is termed the matrix, which is continuous and surrounds the other phase, often called the dispersed phase. The properties of the composites are a function of the properties of the constituent phases, their relative amounts, and the geometry of the dispersed phase. "Dispersed phase geometry" in this context means the shape of the particulates and the particle size, distribution, and orientation. In designing composite materials, scientists, and engineers have ingeniously combined various metals, ceramics, and polymers to produce a new generation of various extraordinary materials.

Over the past decades, compression/injection molded polymer composites have been increasingly used for numerous mechanical and tribological purposes such as seals, gears, and bearings. The feature that makes polymer composites so promising in industrial applications is the possibility of tailoring their properties with functional fillers. It has been found that short fiber reinforcements can generally improve mechanical properties of the polymer composites. Filler reinforced polymer composites can improve the stiffness, decrease thermal expansion, improve long-term mechanical performance and reduce costs. Inorganic fillers are currently used as functional fillers in thermo set/thermoplastic composites.

Glass fiber reinforced polymer matrix composites have been extensively used in various fields such as aerospace industries, automobiles, marine, and defense industries. Their main advantages are good corrosion resistance, lightweight, dielectric characteristic, and better damping characteristics than metals. Incorporating particulate fillers into polymers improves various physical properties of the materials such as mechanical strength, modulus, and heat deflection temperature. The physico-mechanical properties of the polymer matrix composites are determined by the structure of (1) matrix, (2) fiber/filler, and (3) interface.

Epoxy resin represents some of the highest performance resin due to the mechanical properties and resistance to environmental degradation, which leads to their almost exclusive use in aircraft components. Epoxies are defined as cross-linked polymers in which the cross-linking is derived from reactions of the epoxy group. Epoxy resin usually used in coating industry as surface coating materials which combine toughness, flexibility, adhesion, and chemical resistance. In addition, epoxy resin can also be used in both laminating and molding techniques to make fiber reinforcement with better mechanical strength, chemical resistance, and electrical insulating properties [1]. Epoxy resin also used with reinforcing fibers for advanced composites application.

This is due to the capability of epoxy resin that showed good adhesion to the embedded fiber [2].

Epoxy resin as matrix are widely used in the production of glass fiber composites due to their wetting power and adhesion to glass fiber, low setting shrinkage, considerable cohesion strength, adequate dielectric characteristics, and thermal properties. These resins are superior to polyesters in resisting moisture and other environmental influences and offer lower shrinkage and better mechanical properties. However, the structural application of epoxy resin and epoxy resin matrix composites is usually limited owing to the relatively poor thermal stabilities and load carrying capacity. In order to enhance the wear resistance and thermal stabilities, many studies have been carried out. One of these is the modification of matrix. Different fibers and hard particulates made of ceramic or metal particles have been tried as the fillers to modify the epoxy matrix composites for that purpose by several researchers.

Glass fiber is a part of reinforcing materials for reinforced plastics based on single filaments of glass ranging in diameter from 3–19 μm. Glass fibers show good performance and play a main function in playground equipment, recreational items, piping for corrosive chemicals, and many other common applications. The cost of the glass fiber is considerably lower than the cost of carbon-based fibers [3]. Fiber reinforced composites can be explained as the strength of fibers in fibrous composites exceeds considerably the strength of the matrix. Fibers such as short aramid (AF), glass (GF), or carbon (CF) fibers are used in order to increase the creep resistance and the compressive strength of the polymer matrix system used [4]. The addition of fibers is to ensure the strength of the material while the matrix helps to keep the shape of the composite.

The structural application of epoxy resin and epoxy resin matrix composites is usually limited owing to the relatively poor thermal stabilities and load carrying capacity. In order to enhance the wear resistance and thermal stabilities, many studies have been carried out. One of these is the modification of matrix. Different fibers and hard particulates made of ceramic or metal particles have been tried as the fillers to modify the epoxy resin matrix composites for that purpose by several researchers [5-7]. Kim et al. [5] reported that the damage could occur during the fabrication process, storage, service, transport, and maintenance. They are susceptible to mechanical damage when they are subjected to effects of tension, compression, and flexure, which can lead to interlayer delamination. The increase of external load favors the propagation of delamination through the interlayer leading to the catastrophic failure of the component. Another work reported by Unal and Mimaroglu [6] evaluated mechanical properties of Nylon-6 by incorporating one or a combination of more than one filler by varying the weight percent. They observed that the tensile strength and modulus of elasticity of Nylon-6 composites increased with increase in filler weight percent. Varada Rajulu et al. [7] investigated the tensile properties of epoxy toughened with hydroxyl terminated polyester at different layers of glass rovings and reported that the tensile strength increased with increase in fiber content. Their results show that the addition of various fillers/fiber into epoxy matrix enhances the mechanical properties of the composites. The use of such hard particles increases the dry sliding friction coefficient and abrades the counterface. Woven fabric reinforced polymer matrix composites are gaining popularity because of their balanced properties in the fabric plane as well as their ease of handling during fabrication. Also, the simultaneous existence of parallel and anti parallel fibers in a woven configuration leads to a synergetic effect on the enhancement of the wear resistance of the composite [8]. Suresha et al [9] studied the mechanical and sliding wear behavior of SiC filled G-E composites and concluded that the SiC filler addition improved the mechanical as well as wear resistance of G-E composites. Osmani [10] evaluated the mechanical properties of alumina filled G-E composites and reported that tensile and shear strengths decreased with increase in alumina and content and flexural strength and modulus increased. Although there has been considerable research devoted to the mechanical properties of unfilled glass fiber reinforced epoxy resin composites and ceramic or metal particles filled pure epoxy resin composites [5-17], there are no experimental data about the mechanical properties of glass fiber reinforced epoxy composites filled with SiO_2 particles.

Fillers are inert substances added to reduce the resin cost and/or improve its physical properties, viz., hardness, stiffness, and impact strength. Commonly used fillers are calcium carbonate, alumina, graphite, and clay. Incorporating particulate fillers into polymers improves various physical properties of the materials such as mechanical strength, modulus, and heat deflection temperature. The physico-mechanical properties of the polymer matrix composites are determined by the structure of (1) matrix, (2) fiber/filler, and (3) interface.

Hard particulate fillers consisting of ceramic or metal particles and fiber fillers made of glass are being used these days to dramatically improve the properties of composite materials [18]. Various kinds of polymers and polymer matrix composites reinforced with metal particles have a wide range of industrial applications such as secondary load-bearing aerospace structures, circuit boards, ski boats, insulating boards, electrodes materials with thermal durability at high temperature [19, 20] and so on. These engineering composites are desired due to their low density, high corrosion resistance, ease of fabrication, and low cost [21, 22]. Along with fiber-reinforced composites, the composites manufactured with particulate fillers have been found to perform well in many real operational conditions. Currently, particle size is being reduced rapidly and many studies have focused on how single-particle size affects mechanical properties [23]. Yamamoto et al. [24] found that the structure and shape of silica particle have significant effect on the mechanical properties such as fatigue resistance, tensile, and fracture properties. Weizhou Jiao et al. [25] during characterization of mechanical properties of polymeric composites concluded that addition of Al_2O_3 filler can improve the impact and yield strength of polymer composites. Seung-Chul Lee et al [26] reported that quasi-isotropic plies can exhibit excellent flexural behavior compared to cross plies. Nathaniel Chisholm et al [27] found that mechanical properties are declining by increasing the weight% of SiC filler on carbon/epoxy composites. Beckry Abdel-Magid et al. [28] carried out tests on E-glass/epoxy and E-glass/polyurethane and found that E-glass/polyurethane exhibited better mechanical properties compared to E-glass/epoxy. The susceptible mechanical damage occurred when the GFRP components are subjected to effects of tension, compression, and flexure, which can lead to interlayer delamination. The increase of external load favors the propagation of delamination through the interlayer leading to the catastrophic failure of the component. Although there has been considerable research devoted to the mechanical properties of unfilled glass fiber reinforced epoxy resin composites and ceramic or metal particles filled pure epoxy resin composites, there are no experimental data about the mechanical properties of glass fiber reinforced epoxy composites filled with SiO_2 particles. Therefore, an experimental study has been carried out to investigate the physico-mechanical properties of glass fiber reinforced epoxy composite filled with different types and proportions of microfillers.

11.2 MATERIAL USED

Woven glass plain weave fabrics made of 360 g/m² containing E-glass fibers of diameter of about 12 μm have been employed. The epoxy resin (LAPOX L-12) was mixed with the hardener (K-6, supplied by ATUL India Ltd., Gujarat, India) in the ratio 10:1.2 by weight. Eight layers of fabrics were used to obtain approximately laminates of thickness 3 mm. A hand lay-up technique was used to fabricate glass fabric reinforced epoxy composites followed by compression moulding. The weight percent of the glass fiber in the composite is 60. The quality of all laminates was assessed by ultrasonic C-scanning prior to the test.

11.3 FABRICATION OF COMPOSITE LAMINATES

E-glass plain weave woven roving fabric, which is compatible to epoxy resin used as the reinforcement [29]. The epoxy resin is mixed with the hardener in the ratio 100:12 by weight. Dry hand lay-up technique is employed to produce the composites. The stacking procedure consists of placing the fabric one above the other with the resin mix well spread between the fabrics. A porous Teflon film is placed on the completed stack. To ensure uniform thickness of the sample a spacer of size 3 mm is used. The mold plates have a release agent smeared on them. The whole assembly is pressed in a hydraulic press (0.5 MPa) and allowed to cure for a day. The laminate so prepared has a size 250 mm × 250 mm × 3 mm. To prepare the particulate filled G-E composites, selected particulates (average particle size of about 40 μm) are mixed with a known weighed quantity of epoxy resin. The details of the composites made are shown in Table 1.

TABLE 1 Compositional details with sample code of the composites fabricated.

Composites			
Samples name (code)	**Composition by weight %**		
	Epoxy	**Glass fiber**	**Filler**
Glass-epoxy (G-E)	40	60	—
Alumina filled glass-epoxy (5Al$_2$O$_3$-G-E)	35	60	5
Alumina filled glass-epoxy (7.5Al$_2$O$_3$-G-E)	32.5	60	7.5

TABLE 1 *(Continued)*

Composites

Samples name (code)	Composition by weight %		
	Epoxy	Glass fiber	Filler
Graphite filled glass-epoxy (5Gr-G-E)	35	60	5
Graphite filled glass-epoxy (7.5Gr-G-E)	32.5	60	7.5
Slicon carbide filled glass-epoxy (5SiC-G-E)	35	60	5
Slicon carbide filled glass-epoxy (7.5SiC-G-E)	32.5	60	7.5
Slicon dioxide filled glass-epoxy (5SiO$_2$-G-E)	35	60	5
Slicon dioxide filled glass-epoxy (7.5SiO$_2$-G-E)	32.5	60	7.5

11.4 PHYSICO-MECHANICAL TESTS

Density of the composites were determined by using a high precision electronic balance (Mettler Toledo, Model AX 205) using Archimede's principle. Hardness (Shore-D) of the samples was measured as per ASTM D2240, by using a Hiroshima make hardness tester (Durometer). Six readings at different locations were noted and average value is reported. The tensile measurement was carried out using universal tensile testing machine (JJ Lloyd, London, United Kingdom, capacity 1–20 kN), according to ASTM D3039. The tensile test was performed at crosshead speed of 30 mm/min (quasi-static).

The flexural properties, namely, modulus and strength of the particulate filled G-E composites, were determined on rectangular specimens (90 mm × 12 mm × 3 mm) in three points bending at room temperature according to ASTM D790. The span length of the specimens was 70 mm and their loading on a universal testing machine (JJ Lloyd, London, United Kingdom, capacity 1–20 kN) occurred with deformation rate v = 1.3 mm/min.

Izod impact test were carried out using an Avery ceast pendulum impact tester (ASTM D256–92). A 7.5 J/m energy hammer was used and the striking velocity was 3.46 m/s. Five samples were tested for each composite type for all the studies and the average value was recorded.

For select combinations of particulates and unfilled G-E samples are sputter coated with gold for detailed SEM (XL30 SEM with Oxford ISIS310 EDX, England) features.

11.5 DENSITY

Density of a composite depends on the relative proportion of matrix and reinforcing materials and this was one of the most important factors determining the properties of the composites. The measured densities with varied wt% of microfiller in G-E composite are presented in Table 2. It is observed that addition of fillers increased the density of the G-E composites as compared to unfilled one. Type of filler does not seem to influence the density significantly.

11.6 TENSILE BEHAVIOR

The test results of microfiller filled G-E composites for tensile strength and modulus are shown in Table 2. It is seen that for all the composite samples, irrespective of the filler material, the tensile strength of the composite varied with the change in type of filler. The unfilled G-E composite G-E has a tensile strength of 408.16 MPa and it may be seen from Table 2 that the tensile strength value drops from 408.16 to 279.19 MPa with the addition of different microfillers having same wt%. Among the four fillers considered in the present study, the inclusion of SiC causes maximum reduction in the composite strength. There can be two reasons for the decline in the strength of these particulate filled composites compared to the unfilled one. One possibility is that the interfacial adhesion between the filler particles and the matrix may be too weak to transfer the tensile stress: the other is that the sharp corners of the irregularly shaped SiC particles result in stress concentration in the epoxy matrix. The compatibility of SiC particles in epoxy resin seem to be not as good as that of other microfiller; as a result of which the percentage reduction in tensile strength is highest in this case as shown in the Figure 1, Figure 2, Figure 3, and Figure 4.The tensile modulus of filled composites are also found to be less than the modulus of the unfilled G-E.

11.7 FLEXURAL STRENGTH

The static flexural properties that is, flexural modulus, strength, and displacement at maximum load are listed in Table 2. Figure 3, Figure 4, and Figure 7 shows the flexural strengths of G-E and microfiller filled G-E composites. One can establish that the flexural modulus and flexural strength increased, whereas the displacement at maximum load decreased with microparticles incorporation. Accordingly, particulates acted as reinforcement in the related SiC/SiO_2-G-E hybrids. The highest flexural modulus and strength were found for the SiC filled G-E composites.

11.8 IMPACT STRENGTH

The impact energy values of different composites recorded during the impact tests are also given in Table 2. It suggests that the resistance to impact loading of G-E composite improves with addition of particulate fillers except SiC. It is seen that with incorporation of Al_2O_3, graphite, and SiO_2 particles in unfilled G-E composite, the impact strength increased by about 20 to 32%. It is also note worthy that the SiO_2 filled composites show 32% higher impact strength compared to unfilled G-E composite. Figure 8 shows the impact energy of the unfilled and microfiller filled G-E composites.

11.9 HARDNESS (SHORE-D)

The improvement in surface hardness of microfilled G-E composites is not significant as compared with the unfilled G-E composites (Table 2). The surface hardness (Shore-D) values of the unfilled and microfilled composites are in the range 64–69 as shown in Figure 9.

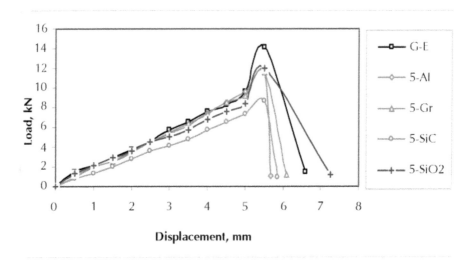

FIGURE 1   Typical load *vs* displacement curves of 5 wt% microfiller filled G-E samples.

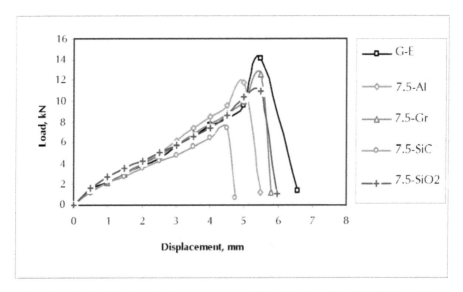

FIGURE 2   Typical load *vs* displacement curves of 7.5 wt% microfiller filled G-E samples.

TABLE 2 Physical and mechanical properties of the composites.

Composites		Density	Tensile strength	Tensile modulus	Flexural strength	Impact energy	Hardness
Sample code	composition	(g/cm³)	(MPa)	(GPa)	(MPa)	(J/m)	(Shore-D)
G-E	(60%GF + 40%E)	1.56	408.16	12.61	399.21	2.63	64
5-Al	(60%GF + 35%E + 5%Al₂O₃)	1.85	328.16	10.96	379.08	3.09	66
7.5-Al	(60%GF + 32.5%E + 7.5%Al₂O₃)	1.89	322.14	11.35	430.24	3.52	69
5-Gr	(60%GF + 35%E + 5%Graphite)	1.88	342.08	11.77	464.13	3.29	68
7.5-Gr	(60%GF + 32.5%E + 7.5%Graphite)	1.85	355.82	11.14	432.22	3.25	67
5-SiC	(60%GF + 35%E + 5% SiC)	1.83	336.06	10.71	602.04	2.31	66
7.5-SiC	(60%GF + 32.5%E + 7.5% SiC)	1.82	279.19	8.82	530.04	2.51	67
5- SiO₂	(60%GF + 35Wt%E + 5% SiO₂)	1.81	403.13	12.45	469.65	3.35	65
7.5-SiO₂	(60%GF+32.5%E + 7.5% SiO₂)	1.79	319.29	9.83	849.26	3.84	66

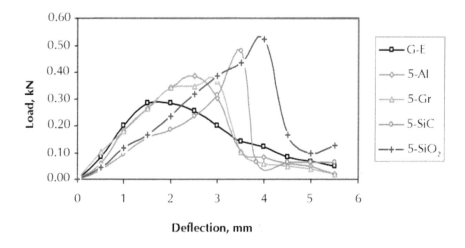

FIGURE 3   Typical load *vs* deflection curves of 5 wt% microfiller filled G-E samples.

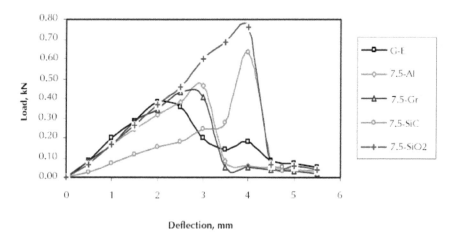

FIGURE 4   Typical load *vs* deflection curves of 7.5 wt% microfiller filled G-E samples.

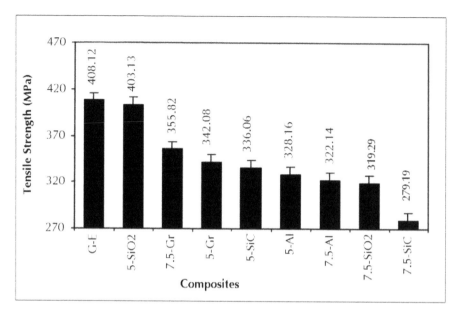

FIGURE 5   Tensile strength of unfilled and microfiller filled G-E composites.

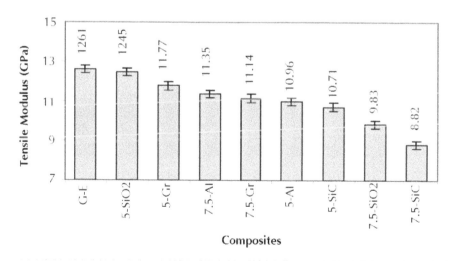

FIGURE 6   Tensile modulus of unfilled and microfiller filled G-E composites.

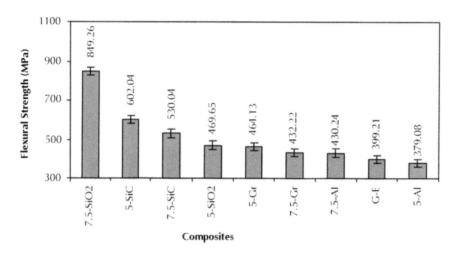

FIGURE 7   Flexural strength of unfilled and microfiller filled G-E composites.

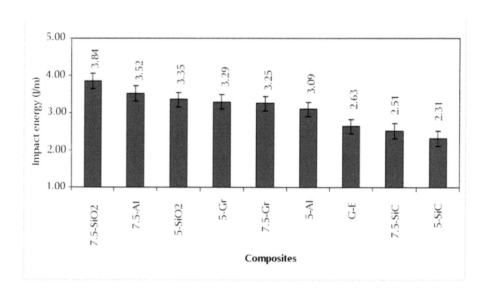

FIGURE 8   Impact energy of unfilled and microfiller filled G-E composites.

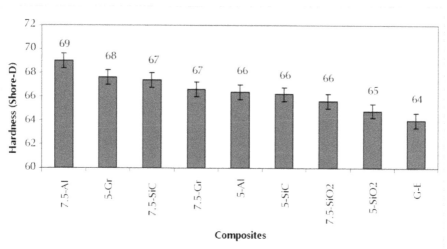

FIGURE 9   Shore-D Hardness of unfilled and microfiller filled G-E composites.

11.10   FRACTOGRAPHIC ANALYSIS

The SEM micrographs in Figure 10(a) and (b), Figure 11(a) and (b), and Figure 12(a) and (b) show the tensile fractured surface of G-E, SiC-G-E, and SiO_2 filled G-E composites respectively. The photomicrographs revealed the linear elastic behavior and brittle type fracture for the test samples along with instant multiple fractures. The fracture is due to delamination between the layers of the composite samples and fiber pulls out (Figure 10(a)). In the G-E composite ductile elongation is observed (Table 2). It is interesting to note that the composite characterized by higher tensile strength showed more brittle fractures. For G-E sample, the fracture is ductile-brittle and can be explained by the plastic deformation of the matrix after fiber matrix debonding. The SEM micrograph shown in Figure 10(b) supports this failure mechanism because the fibers on fractured surfaces are clean, which shows brittle fracture. Other important failure mechanism of composites such as fiber matrix debonding (marked 'a'), fiber fracture (marked 'b'), and cohesive resin fracture (marked 'c') are also observed in a SEM microphotograph (Figure 10(b)). Generally, matrix fracture was found to initiate at the surface of the fibers as indicated by the direction of river lines (marked by an arrow) and propagates into the resin on either side, where cracks extend from the surfaces of adjacent fibers simultaneously [30].

The SEM characterization of the SiC-G-E fractured surface shows (Figure 11(a) and (b)) that the fibers are more or less covered with the matrix and SiC particles (marked by arrows) a qualitative indication of a greater interfacial strength. Disorientation of transverse fibers (marked 'P' in Figure 11(a)), fiber bridging (marked 'Q' in Figure 11(a)), fibers pull out, inclined fracture of longitudinal fibers (marked 'R' in Figure 11(b)), matrix rollers, and matrix cracking (marked 'S' in Figure 11(b)) is also seen. The improvement reported in terms of mechanical properties of the composites evaluated is mainly due to the enhancement of adhesion or interfacial interactions among the fibers, matrix, and SiC filler [31].

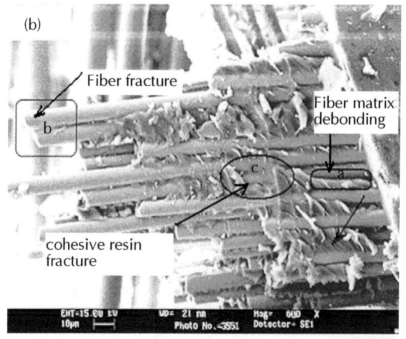

FIGURE 10   Tensile fractured surfaces of unfilled G-E composites [30].

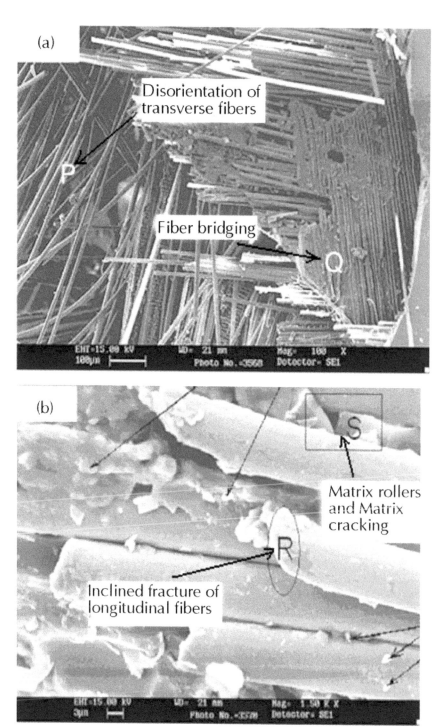

FIGURE 11 Tensile fractured surfaces of SiC filled G-E composites [30].

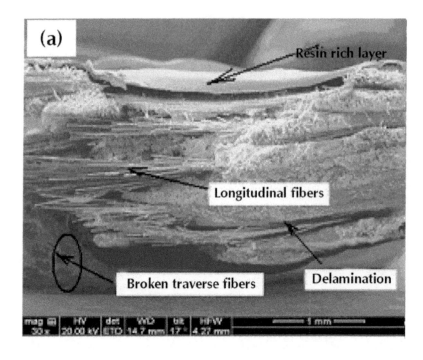

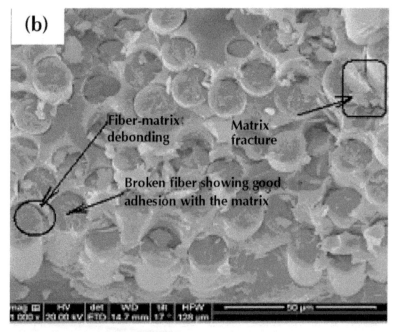

FIGURE 12   Tensile fractured surfaces of SiO$_2$ filled G-E composites.

The SEM characterization of the SiO_2-G-E fractured surface shows (Figures 12(a) and (b)) that the fibers as well as SiO_2 particles are adhered to the matrix very well (Figure 12(b)) and a qualitative indication of a greater interfacial strength. Disorientation of transverse fibers (Figure 12(a)), fiber bridging, fibers pull out (Figure 12(a)), and normal fracture of longitudinal fibers (Figure 12(b)), matrix debris, and matrix cracking (Figure 12(b)) is also seen. The improvement reported in terms of mechanical properties of the composites evaluated is mainly due to the enhancement of adhesion or interfacial interactions among the fibers, matrix, and SiO_2 filler.

11.11 CONCLUSION

In this study, an experimental investigation has been conducted to evaluate the mechanical properties of glass fiber reinforced epoxy matrix filled with different proportions of very fine SiC, SiO_2, graphite, and Al_2O_3 particles. The present work showed that it is possible to successfully fabricate a multi component hybrid composite (using epoxy as matrix, glass fiber as reinforcement and alumina, graphite. The SiC, SiO_2 as filler) by hand lay-up technique. The following main conclusion can be drawn from this study.

The work showed that incorporation of SiC and SiO_2 fillers in G-E composites modifies the tensile, flexure and impact strength of the composites. The ultimate tensile strength and Young's modulus of the glass fiber reinforced epoxy resin composites decreased effectively with increasing the SiC and SiO_2 filler loadings. The flexural properties of the glass fiber reinforced epoxy composites increased with increasing SiC and SiO_2 particle content. The flexural properties showed an increasing trend with addition of SiC and SiO_2 fillers in G-E composites. In order to reach the best flexural properties, the optimal content of SiO_2 particles in the filled epoxy matrix is recommended as 7.5 wt% compared with the flexural properties of the unfilled glass fiber reinforced epoxy composite, with the addition of 7.5 wt% of SiO_2 particle in the matrix, flexural strength and flexural modulus were increased by 112 and 113%, respectively. The hardness and density of the composites are also influenced by the type and content of microfillers.

The author believed that the sea water uptake deteriorated the mechanical properties of unfilled and filled G-E composites. Disorientation of transverse fibers, fiber bridging, fibers pull out, inclined as well as normal fracture of longitudinal fibers, matrix rollers, and matrix cracking are the fracture mechanisms during the tensile test.

KEYWORDS

- **Dispersed phase geometry**
- **Epoxy resin**
- **Flexural strength**
- **Hand lay-up technique**
- **Microfiller**

REFERENCES

1. Wilkinson, S. B. and White, J. R. Thermosetting short fiber reinforced composites. In *Short fiber-polymer composites*. S. K De and J. R. White. (Eds.), Woodhead Publishing Limited, Cambridge, pp. 54–81 (1996).
2. Zhou, Y., Pervin, F., Biswas, M. A., Rangari, V. K., and Jeelani, S. Fabrication and characterization of montmorillonite clay-filled SC-15 epoxy. *Materials Letters*, **60**, 869–873 (2006).
3. Hyer, M. W. *Stress analysis of fiber-reinforced composite materials*. McGraw-Hill, USA, pp. 578–699 (1998).
4. Friedrich, K., Zhang, Z., and Schlarb, A. K. Effects of various fillers on the sliding wear of polymer composites. *Composites Science and Technology*, **65**, 2329–2343 (2005).
5. Kim, J., Shioya, M., Kobayashi, H., Kaneko, J., and Kido, M. Mechanical Properties of Woven Laminates and Felt Composites using Carbon Fibers, Part 1: Enhancement of Mechanical Properties. *Composite Science and Technology*, **64**(13–14), 2221–2229 (2004).
6. Unal, H. and Mimaroglu, A. Influence of Filler Addition on the Mechanical Properties of Nylon-6 Polymer. *Journal of Reinforced Plastics and Composites*, **23**(5) 461–469 (2004).
7. Varada Rajulu, A., Sanjeev Kumar, S. V., Babu Rao, G., Shashidhara, G. M., He Song, and Jun Zhang. Tensile Properties of Glass Rovings/Hydroxyl Terminated Polyester Toughened Epoxy Composites. *Journal of Reinforced Plastics and Composites*, **21**(17) 1591–1596 (2002).
8. Viswanath, B., Verma, A. P., and Kameswara Rao, C. V. S. Effect of Matrix Content on Strength and Wear of Woven Roving Glass Polymeric Composites. *Composite Science and Technology*, **44**(2), 77–86 (1992).
9. Suresha, B., Chandramohan, G., Sadananda Rao, P. R., Samapthkumaran, P., and Seetharamu, S. Influence of SiC Filler on Mechanical and Tribological Behavior of Glass Fabric Reinforced Epoxy Composite Systems. *Journal of Reinforced Plastics and Composites*, **26**, 565–578 (2007).
10. Osmani, A. Mechanical Properties of Glass Fiber Reinforced Epoxy Composites Filled with Al_2O_3 Particles. *Journal of Reinforced Plastics and Composites*, **28**, 2861–2867 (2009).
11. Hussain, M., Oku, Y., Nakahira, A., and Niihara, K. Effects of Wet Ball-milling on Particle Dispersion and Mechanical Properties of Particulate Epoxy Composites. *Materials Letters*, **26**, 177–184 (1996).
12. Wetzel, B., Haupert, F., and Zhang, M. Q. Epoxy Nanocomposites with High Mechanical and Tribological Performance. *Composites Science and Technology*, **63**, 2055–2067 (2003).
13. Clements, L. L. and Moore, R. L. Composite Properties for E-glass Fibers in a Room Temperature Curable Epoxy Matrix. *Composites*, 93–99 (April, 1978).
14. Shindo, Y., Takano, S., Narita, F., and Horiguchi, K. Tensile and Damage Behavior of Plain Weave Glass/Epoxy Composites at Cryogenic Temperatures. *Fusion Engineering and Design*, **81**, 2479–2483 (2006).
15. Gu, H. Tensile and Bending Behaviors of Laminates with Various Fabric Orientations. *Materials and Design*, **27**, 1086–1089 (2006).
16. Zhao, S., Zhang, J., Zhao, S., Li, W., and Li, H. Effect of Inorganic-Organic Interface Adhesion on Mechanical Properties of Al_2O_3/Polymer Laminate Composites. *Composite Science and Technology*, **63**, 1009–1014 (2003).
17. Wetzel, B., Rosso, P., Haupert, F., and Friedrich, K. Epoxy Nanocomposites-fracture and Toughening Mechanisms. *Engineering Fracture Mechanics*, **73**, 2375–2398 (2006).
18. Gregory, S. W., Freudenberg, K. D., Bhimaraj, P., and Schadler, L. S. A study on the friction and wear behavior of PTFE filled with alumina nanoparticles. *J. Wear*, **254**, 573–580 (2003).
19. Jung-Il, K., Kang P. H., and Nho, Y. C. Positive temperature coefficient behavior of polymer composites having a high melting temperature. *J. Appl. Poly Sci.*, **92**, 394–401 (2004).
20. Zhu, K. and Schmauder, S. Prediction of the failure properties of short fiber reinforced composites with metal and polymer matrix. *J. Comput. Mater. Sci.*, **28**, 743–748 (2003).
21. Rusu, M., Sofian, N., and Rusu, D. Mechanical and thermal properties of zinc powder filled high density polyethylene composites. *J. Polymer Testing*, **20**, 409–417 (2001).

22. Tavman, I. H. Thermal and mechanical properties of copper powder filled poly ethylene) composites. *J. Powder Tech.*, **91**, 63–67 (1997).
23. Cantwell, W. J. and Moloney, A. C. *Fractography and failure mechanisms of polymers and composites*. Elesvier, Amsterdam, p. 233 (1994).
24. Yamamoto, I., Higashihara, T., and Kobayashi, T. Effect of silica-particle characteristics on impact/usual fatigue properties and evaluation of mechanical characteristics of silica-particle epoxy resins. *Int. J. JSME*, **46**(2), 145–153 (2003).
25. Weizhou Jiao, Youzhi Liu, and Guisheng Qi. Studies on mechanical properties of epoxy composites filled with the grafted particles PGMA/Alumina. *Composite science and Technology*, **69**, 391–395 (2009).
26. Seung-Chul Lee, Seong-Taek Jeong, Jong-Nam Park, Sun Jin Kim, and Gyu-Jae Cho. A study on mechanical properties of carbon fiber reinforced plastics by three-point bending testing and transverse static response. *Journal of material processing technology*, **201**, 761–764 (2008).
27. Nathaniel Chisholm, Hassan Mahfuz, Vijaya Rangari, Reneé Rodgers, and Shaik eelani. Synthesis and Mechanical Characterization of Carbon/Epoxy Composites Reinforced with SiC Nanoparticles. *NSTI-Nanotech 2004*, www.nsti.org, ISBN 0-9728422-9-2 Vol. 3, p. 302 (2004).
28. Beckry Abdel-Magid, Roberto Lopez-Anido, Glenn Smith, Sara Trofka. Flexure creep properties of E-glass reinforced polymers. *Composite Structures*, **62**, 247–253 (2003).
29. Lee, H. and Nevilee, K. *Handbook of Epoxy Resins*. McGraw-Hill, New York (1967).
30. Suresha, B., Chandramohan, G., Renukappa, N. M., and Siddaramaiah. Influence of silicon carbide filler on mechanical and dielectric properties of glass fabric reinforced epoxy composites. *J Appl Poly Sci.* **111**, 685–91 (2009).
31. Suresha, B., Chandramohan, G., Kishore, P., Samapathkumaran, and Seetharamu, S. Mechanical and three-body abrasive wear behavior of SiC filled glass-epoxy composites. *Polym Compos.*, **29**, 1020–1025 (2008).

12 Corrosion Performance of Heat Treated Ni–CO Nanocomposite Coatings

B. Ranjith and G. Paruthimal Kalaignan

CONTENTS

12.1 INTRODUCTION

Hot corrosion is a serious problem in power generation equipment, gas turbines for ships and aircraft, and in other energy conversion and chemical process systems. No alloy is immune to hot corrosion attack indefinitely. Super alloys have been developed for high temperature applications. However, these alloys may not be able to meet both the high temperature strength requirements and high temperature corrosion resistance simultaneously for longer life. So, protective coatings are used to counter the latter. Due to continuously rising cost of the materials as well as increased material requirements, the coating techniques have been given more importance in recent times. Coatings can add value to products up to ten times the cost of the coating [1]. The necessity to preserve good mechanical properties of alloys at elevated temperature under the highly oxidizing and corrosive conditions led to the development of coating materials [2, 3]. The composition and structure of the coatings are determined by the role that, they play in various material systems and performance environments and may vary from one system to another according to the requirements. In the service environment, the coatings are expected to form protective oxides and carbides such as TiO_2, Al_2O_3,

Cr_2O_3, SiC, and WC. So, the coatings are designed to serve as a reservoir for the elements forming or contributing to form these surface oxides and carbides. Nickel-based alloy coatings show good high temperature wear and corrosion resistance. They have good wear resistance after adding W and Mo elements to the alloy [4]. Nickel-based coatings are used in applications when wear resistance combined with oxidation or hot corrosion resistance is required. Nickel-based self-fluxing alloys are mainly used in the chemical industry, petrol industry, glass mould industry, hot working punches, fan blades, and mud purging elements in cement factories. Their advantages are especially related to coating large-sized components such as piston rods, earth working machines, and so on [5].

The Ni–Co–TiO_2/SiC and WC coated plates (plated from optimum composition) were subjected to heat treatment at three different temperatures viz 700, 800, and 900°C using muffle furnace. The heat treated plates were subsequently tested for corrosion resistance using Tafel polarization and electrochemical impedance spectroscopy. Particularly, Ni–Co–TiO_2 (6 g/l), Ni–Co–SiC (8 g/l) and Ni–Co–WC (4 g/l) composite coated on mild steel plate at 900°C is much lower corrosion potential (–0.4977 V and –0.4670 V) compared to other temperatures 700 and 800°C. It reveals that, the Ni–Co–TiO_2 (6 g/l), Ni–Co–SiC (8 g/l), and Ni–Co–WC (4 g/l) nanocomposite coated mild steel plate at 900°C protect the corrosion reaction of mild steel plate in 3.5% NaCl solution. The increase in R_{ct} values and decrease in C_{dl} values were indicated the good corrosion inhibition.

12.2   EXPERIMENTAL DETAILS

The following nanocomposites viz Ni–Co–TiO_2 (6 g/l), Ni–Co–SiC (8 g/l), and Ni–Co–WC (4 g/l) coatings are prepared by PC method. The Ni–Co–TiO_2, SiC and WC coated plates (plated from optimum composition) were subjected to heat treatment at three different temperatures viz 700, 800, and 900°C using hot air oven. The heat treated plates were subsequently tested for corrosion resistance using Tafel polarization and electrochemical impedance spectroscopy.

12.3   DISCUSSION AND RESULTS

12.3.1   Tafel Polarization Curves For Ni–Co Nanocomposites Coated Specimens

The electrochemical parameters obtained from the polarization curves for Ni–Co–TiO_2 (6 g/l), Ni–Co–SiC (8 g/l) and Ni–Co–WC (4 g/l) composite coated on mild steel plates at 700, 800, and 900°C were given in Table 1. Figures 1, 2, and 3 presents the potentiodynamic curves for Ni–Co–TiO_2 (6 g/l), Ni–Co–SiC (8 g/l), and Ni–Co–WC (4 g/l) composite coated on mild steel plates at 700, 800, and 900°C.

Particularly Ni–Co–TiO_2 (6 g/l), Ni–Co–SiC (8 g/l), and Ni–Co–WC (4 g/l) composite coated on mild steel plate at 900°C is much lower corrosion potential (–0.4977 and–0.4670 V) compared to other temperatures 700 and 800°C which reveals that, Ni–Co–TiO_2 (6 g/l), Ni–Co–SiC (8 g/l), and Ni–Co–WC (4 g/l) nanocomposite at 900°C is excellent corrosion protectant for mild steel plate. The corrosion current density (I_{corr}) of Ni–Co–TiO_2 (6 g/l), Ni–Co–SiC (8 g/l), and Ni–Co–WC (4 g/l) nanocomposite coated on mild steel plate at 900°C has much lower I_{corr} value than other temperatures 700 and 800°C.

TABLE 1 Parameters derived from polarization curves for Ni–Co–TiO$_2$ (6 g/l), Ni–Co–SiC (8 g/l), and Ni–Co–WC (4 g/l) nanocomposites coated on mild steel plates in 3.5% NaCl.

S. No	Samples	Temperature (°C)	E_{corr} (V)	I_{corr} (10^{-5}A/cm^2)
		700	−0.5504	3.780
1.	Ni–C0–TiO$_2$ (6 g/l)	800	−0.5304	3.222
		900	−0.4977	3.150
		700	−0.5754	3.102
2.	Ni–Co–SiC (8 g/l)	800	−0.4891	2.299
		900	−0.4670	2.249
		700	−0.5642	3.623
3.	Ni–Co –WC (4 g/l)	800	−0.5052	3.415
		900	−0.4897	3.163

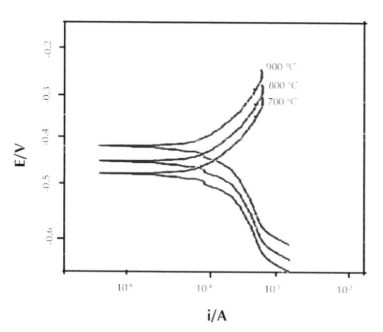

FIGURE 1 Potentiodynamic polarization curves for Ni–Co–TiO$_2$ (6 g/l) nanocomposite coatings at different temperatures.

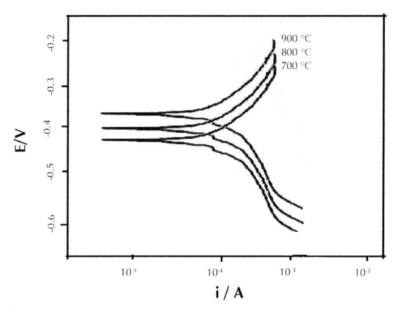

FIGURE 2 Potentiodynamic polarization curves for Ni–Co–SiC (8 g/l) nanocomposite coatings at different Temperatures.

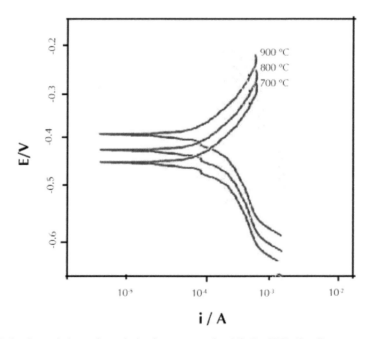

FIGURE 3 Potentiodynamic polarization curves for Ni–Co–WC (8 g/l) nanocomposite coatsings at different temperatures.

12.3.2 Electrochemical Impedance Spectroscopy

The corrosion protection of mild steel plate by Ni–Co–TiO$_2$ (6 g/l), Ni–Co–SiC (8 g/l), and Ni–Co–WC (4 g/l) composite coated on Mild steel plates at 700, 800, and 900°C were investigated by electrochemical impedance spectroscopy. The Nyquist plot of Ni–Co–TiO$_2$ (6 g/l), Ni–Co–SiC (8 g/l), and Ni–Co–WC (4 g/l) nanocomposite coated on mild steel plates at 700, 800, and 900°C is shown in Figures 4, 5, and 6. The parameters derived from Nyquist plots and the corresponding data's are represented in Table 2.

TABLE 2 Parameters derived from polarization curves for Ni–Co–TiO$_2$ (4 g/l), Ni–Co–SiC (8 g/l) and Ni–Co–WC (4 g/l) nanocomposites Coated on mild steel plates in 3.5% NaCl.

S. No	Samples	Temperature (^{0} C)	R$_{ct}$ (ohms)	C$_{dl}$ (Fcm^{-2}) X 10^{-5}
1.	Ni–C0–TiO$_2$ (6 g/l)	700	313	4.25
		800	321	4.18
		900	356	4.09
2.	Ni–Co–SiC (8 g/l)	700	319	4.19
		800	324	4.15
		900	341	4.07
3.	Ni–Co–WC (4 g/l)	700	320	4.13
		800	329	4.10
		900	349	4.06

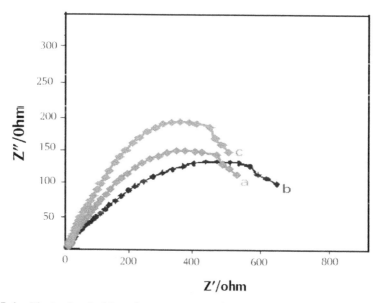

FIGURE 4 Electrochemical impedance curves of Ni–Co–TiO (6 g/l) nanocomposite at (a) 700°C (b) 800°C (c) 900 0 C coating samples.

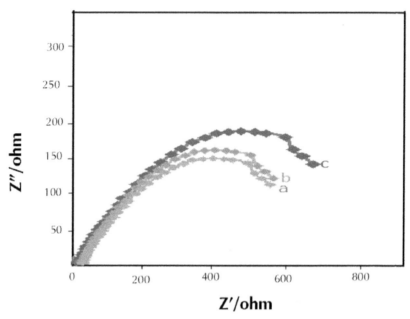

FIGURE 5 Electrochemical impedance curves of Ni–Co–SiC (8 g/l) nanocomposite at (a) 700°C (b) 800°C (c) 900°C coating samples.

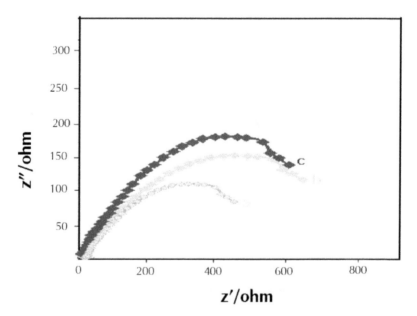

FIGURE 6 Electrochemical impedance curves of Ni–Co–WC (8 g/l) nanocomposite at (a) 700°C (b) 800°C (c) 900°C coating samples.

The Ni–Co–TiO$_2$ (6 g/l), Ni–Co–SiC (8 g/l), and Ni–Co– WC (4 g/l) nanocomposite coated mild steel plate at 900°C has higher R$_{ct}$ value than other Ni–Co–TiO$_2$ (6 g/l), Ni–Co–SiC (8 g/l) and Ni–Co–WC (4 g/l) nanocomposite coated at 700 and 800°C. It reveals that, the Ni–Co–TiO$_2$ (6 g/l), Ni–Co–SiC (8 g/l), and Ni–Co–WC (4 g/l) nanocomposite coated mild steel plate at 900°C protect the corrosion reaction of mild steel plate in 3.5% NaCl solution. The increase in R$_{ct}$ values and decrease in C$_{dl}$ values were indicated the good corrosion inhibition.

12.4 CONCLUSION

The Ni–Co–TiO$_2$ (6 g/l), Ni–Co–SiC (8 g/l), and Ni–Co–WC (4 g/l) composite coated on mild steel plate heat treated at 900°C showed much lower corrosion potential (–0.4977 and–0.4670 V) compared to other temperatures 700 and 800°C which reveals that Ni–Co–TiO$_2$ (6 g/l), Ni–Co–SiC (8 g/l), and Ni–Co–WC (4 g/l) nanocomposite at 900°C is excellent corrosion protectant for mild steel plate. The corrosion current density (I$_{corr}$) of Ni–Co–TiO$_2$ (6 g/l), Ni–Co–SiC (8 g/l), and Ni–Co–WC (4 g/l) nanocomposite coated on mild steel plate at 900°C has much lower I$_{corr}$ value than other temperatures 700 and 800°C.

KEYWORDS

- **Corrosion performance**
- **High temperature performance**
- **Hot corrosion**
- **Nyquist plots**
- **Tafel and EIS**

REFERENCES

1. Matthews, A., Artley, R. J., and Holiday, P. Future's bright for surface engineering. *Mater. World*, **6**, 346–347 (1998).
2. Wood, J. H. and Goldman, E. H. In *Superalloys II*. C. T. Sims, N. S. Stoloff, and W. C. Hagel. (Eds.). Wiley, New York, p. 359 (1987).
3. Liu, P. S., Liang, K. M., and Gu, S. R. High-temperature oxidation behavior of aluminide coatings on a new cobalt-base superalloy in air. *Corros. Sci.*, **43**, 1217–1226 (2001).
4. Rosso, M. and Bennani, A. Studies of new applications of Nickel-based powders for hardfacing processes. *PM World Congress Thermal Spraying/Spray Forming*, 524–530 (1998).
5. Tan, J. C., Looney, L., and Hashmi, M. S. J. Component repair using HVOF thermal spraying. *JMPT*, **92–93**, 203–208 (1999).

13 Field Emission Study of Thermally Exfoliated Wrinkled Graphene

Tessy Theres Baby and Sundara Ramaprabhu

CONTENTS

13.1 INTRODUCTION

The two dimensional (2D) graphene is the rising star in the field of nanotechnology among scientific community because of its exceptional properties. Single layer of graphene is having peculiar properties like high electrical and thermal conductivity, good ballistic transport, chemical inertness, and so on [1, 2]. In 2004, Geim and Novoslev have successfully extracted and characterized single layer of graphene by micro mechanical cleavage [3]. After that, different techniques have been developed for small scale and large scales synthesize of single layer and few layer (2-5 layer) graphene. The different methods of synthesize of graphene include thermal exfoliation [4], hydrogen induced exfoliation of graphite oxide (GO) [5], chemical vapor deposition [6], and so on. Depending on the way in which graphene synthesis, the numbers of layers as well as properties of graphene change.

Electron sources are getting tremendous applications in modern society and play a central role in flat panel displays. The power of graphene as electron field emitters are already been reported with considerably low turn on field and high current density [7]. Field emission involves the extraction of electrons from a metal by tunneling through the surface potential barrier. The emitted current directly depends on the local electric field at the emitting surface E, and its work function φ by the famous Fowler–Nordheim (FN) relation:

$$I = (aAE_{local}^2 / \varphi)\exp(-B\varphi^{\frac{3}{2}} / E_{local}) \tag{1}$$

where $a = 1.54 \times 10^{-6}$ A eV V^{-2} and $B = 6.83 \times 10^7$ eV$^{3/2}$ V cm^{-1}, respectively. A is emission area of the field emitter. E_{local} is the localized electric field, $E_{local} = \beta\dfrac{V}{d} = \beta E$, where β is the field enhancement factor, $\beta = (B\varphi^{3/2}\text{d})/(slope\ of\ FN\ plot)$ and 'd' is the distance between cathode and anode. The present work describes the field emission property of wrinkled graphene synthesized by thermal exfoliation of GO.

Carbon based nanomaterials are considered to be good candidates for field emission applications because of their high aspect ratio, high conductivity, and mechanical strength. Recently, 2D graphene got considerable importance and interest for field emission studies suggest that the wrinkled morphology of exfoliated graphene improves the field emission property of graphene since the wrinkled nature helps for fast electron emission. In the present study, we have synthesized graphene by thermal exfoliation of GO at 1,050°C in Ar atmosphere and studied the field emission properties under high vacuum. The graphene is characterized by different experimental techniques like X-ray diffraction (XRD), BET surface area, atomic force microscopy, scanning electron microscopy and transmission electron microscopy TEM. The field emission study is carried out in an indigenously fabricated setup under a vacuum of ~10^{-6} mbar. The emission behavior is analyzed using the famous FN equation. The fabricated graphene based field emitter showed low turn-on field and high emission current density.

13.2   MATERIALS

The synthesis of graphene is a two step method. In the first step, GO was prepared by hummers method [8]. In the next step, graphene was synthesized by treating vacuum dried GO in Argon gas atmosphere at 1,050°C for 30 s. This graphene is called thermally exfoliated graphene (TEG). Ten milligram of TEG is dispersed in 1ml of 0.5% Nafion solution by ultrasonication. Nafion is used to get a proper dispersion of TEG and also for the good adhesion of TEG with carbon cloth. This dispersion is later spin coated on a flexible carbon cloth using 500 rpm in the first stage and 2,000 rpm in the second stage. The film is heated under vacuum for 12 hr. An indigenously fabricated setup connected to a high vacuum system is used for testing the field emission property of the sample. A mica sheet with 1 mm diameter hole is used as a separator between the cathode and the anode.

Powder XRD studies were carried out using a PAN alytical X'PERT Pro X-ray diffractometer with nickel filtered Cu K$_a$ radiation as the X-ray source. The pattern was recorded in the 2θ range of 5–90° with a step size of 0.016°. Brunauer-Emmett-Teller (BET) surface area analysis was determined by recording nitrogen adsorption/desorption isotherms at 77K using a static volumetric technique with a Micromeritics ASAP V3.01 G 2020 instrument. Field emission scanning electron microscopy (FESEM) and TEM images were taken using FESEM, FEI QUANTA, and JEOL TEM-2010F instruments respectively. Surface roughness of the HEG-200 was obtained by using atomic force microscope (AFM, Dimension 3100, Nanoscope IV Digital Instruments, USA) in tapping mode. The sample for AFM was prepared by dispersing the graphene powder in N-methyl-2-pyrrolidinone (NMP) by ultrasonication and spin coated on a cleaned glass substrate. The sample was then heated to remove the solvent. Field emission study has been done in an indigenously fabricated setup under a pressure of ~10^{-6} mbar [9].

13.3 DISCUSSION AND RESULTS

Figure 1(a) shows the XRD pattern of as synthesized TEG. The broad peak centered around 24° corresponds to the (002) plane of hexagonal graphite (peak ~ 26.5°) with an interlayer spacing of about 0.37 nm. The broad peak is an indication of the short range order of the material. The BET surface area measurement has been done for as grown powdered TEG and is shown in Figure 1(b). The calculated BET surface area from the linear plot is ~470 m²/g. The surface morphology of the as synthesized TEG (Figure 2(a)) and spin coated TEG on carbon cloth (Figure 2(b)) is shown in the FESEM image. The wrinkled morphology of as synthesized TEG is visible in the powder FESEM image. After coating over carbon cloth, we did not observe any change in the morphology. The surface morphology is clearer from TEM image of TEG shown in Figure 2(c).

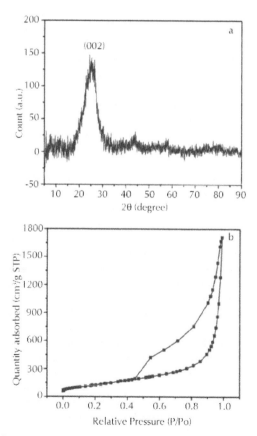

FIGURE 1 (a) XRD pattern and (b) BET surface area measurement of TEG.

FIGURE 2 *(Continued)*

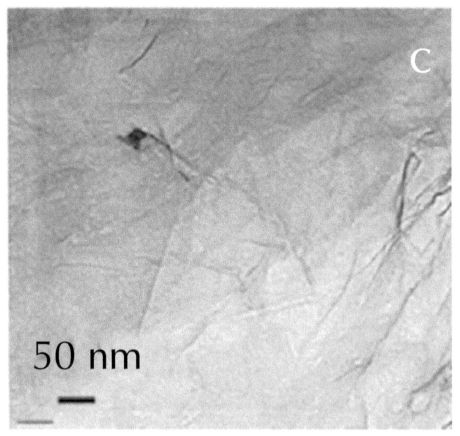

FIGURE 2   The FESEM image of (a) powder TEG, (b) TEG film, and (c) TEM image of TEG.

The J-E plot for fabricated TEG field emitter is shown in Figure 3(a). The turn-on fields for a current density of 10 $\mu A/cm^2$ and threshold field for a current density of 0.2 mA/cm^2 are ~1.82 and ~1.92 V/μm, respectively. The field enhancement factor β has been calculated from the Fowler-Nordheim plot and is around 3,677. The turn-on field reported for a planar graphene is 12.1V/μm [10]. In the case of planar graphene the emission is only from the edges. In the present case, the low turn-on field and threshold field can be due to the wrinkled morphology of TEG, which gives more sites for electron emission. This has been reflected in the field enhancement factor. The wrinkled morphology and surface roughness has been confirmed from AFM image (Figure 4(a)) and corresponding thickness profile (Figure 4(b)). It is not possible to find out number of layer for a wrinkled graphene from AFM. The current stability of the sample has been monitored continuously for a time period of 4 hr for a current density of 0.2 mA/cm^2. The corresponding stability curve is shown in Figure 5. The emission current remained fairly constant for the fabricated field emitter. The fluctuation in emission current/voltage for TEG field emitter is within 4%. Experiment is repeated several times to study the repeatability of measurement.

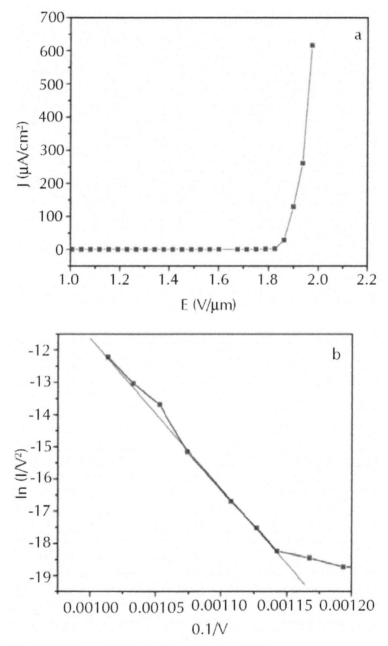

FIGURE 3 (a) J-E plot and (b) Fowler-Nordheim plot of TEG.

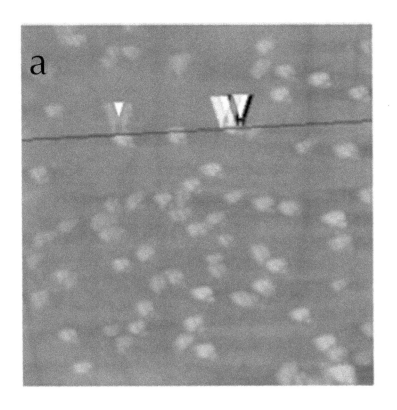

Section Analysis

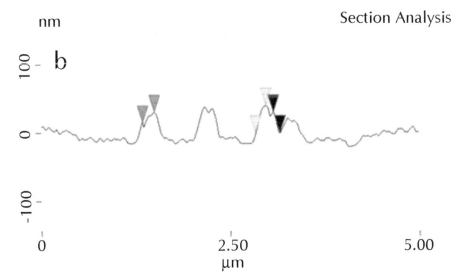

FIGURE 4   (a) AFM image of TEG and (b) corresponding thickness profile.

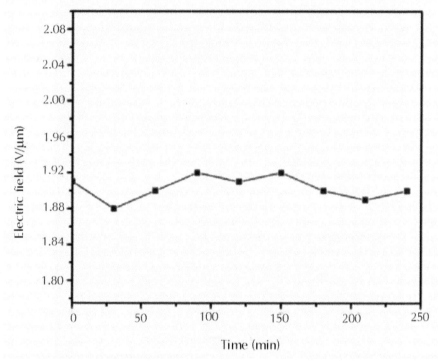

FIGURE 5   Stability of the TEG field emitter has been investigated for a time period of 4 hr.

13.4   CONCLUSION

The TEG was synthesized and characterized by different experimental techniques. The field emission study of TEG coated carbon cloth was studied using an indigenously fabricated setup. The low turn on and threshold voltage is due to the high wrinkle morphology of graphene. Since the sample is coated on a flexible substrate the present field emitter can be used for flexible flat panel displays.

KEYWORDS

- **Field emission**
- **Thermal exfoliation,**
- **Turn-on field**
- **Wrinkled graphenes**

ACKNOWLEDGMENT

The authors wish to thank Indian Institute of Technology Madras, India for the financial support.

REFERENCES

1. Chen, J. H., Jang, C., Xiao, S., Ishigami, M., and Fuhrer, M. S. *Nature Nanotech.*, **3**, 206–209 (2008).
2. Geim, A. K. and Kim, P. *Scientific American*, **298**, 90–97 (2008).
3. Novoselov, K. S., Geim, A. K., Morozov, S. V., Jiang, D., Zhang, Y., Dubonos, S. V., Grigorieva, I. V., and Firsov, A. A. *Science*, **306**, 666–669 (2004).
4. Schniepp, H. C., Li, J. L., McAllister, M. J., Sai, H., Alonso, M. H., Adamson, D. H., Prud'homme, R. K., Car, R., Saville, D. A., and Aksay, I. A. *J. Phys. Chem. B*, **110**, 8535–8539 (2006).
5. Kaniyoor, A., Baby, T. T., and Ramaprabhu, S. *J. Mater. Chem.*, **20**, 8467–8469 (2010).
6. Somani, P. R., Somani, S. P., and Umeno, M. *Chem. Phy. Lett.*, **430**, 56–59 (2006).
7. Eda, G., Unalan, H. E., Rupesinghe, N., Amaratunga, G. A. J., and Chhowalla, M. *Appl. Phy. Lett.*, **93**, 233502-233503 (2008).
8. Hummers, W. S. and Offeman, R. E. *J. Am. Chem. Soc.* **80**, 1339 (1958).
9. Rakhi, R. B., Sethupathi, K., and Ramaprabhu, S. *Nanoscale Res. Lett.*, **2**, 331–336 (2007).
10. Lee, S. W., Lee, S. S., and Yang, E. H. *Nanoscale Res. Lett.* **4**, 1218–1221 (2009).

14 Synthesis and Characterizations of Anatase TiO$_2$ Nanoparticles by Simple Polymer Gel Technique

Prathibha Vasudevan, Sunil Thomas,
Siby Mathew, and N. V. Unnikrishnan

CONTENTS

14.1 INTRODUCTION

Titanium dioxide (TiO$_2$) has been used extensively in the environmental protection procedures: for example, air purification and water disinfection [1]. The properties of TiO$_2$, especially its photocatalytic properties, have proved to be strongly related to its crystal structure, morphology, and crystallite size. TiO$_2$ is known to crystallize as rutile (tetragonal), anatase (tetragonal), and brookite (orthorombic) phases [2]. Among them, TiO$_2$ exists mostly as anatase and rutile phases with a tetragonal structure. Anatase TiO$_2$ has a superior photocatalytic ability compared to rutile TiO$_2$ and poly crystalline TiO$_2$ is superior to amorphous TiO$_2$ [3]. The properties of TiO$_2$ depend upon its structure and form. In comparison to other crystalline phases of TiO$_2$, such as rutile and brookite, the most photoactive phase of TiO$_2$ is anatase. The presence of brookite phase along with anatase and the conversion of anatase to rutile phase at high temperatures are the major challenging factors for the synthesis of pure anatase TiO$_2$. Here we

present a simple and cost effective method of producing nanocrystalline and pure anatase TiO_2 in powder form at lower temperature by polymer gel technique. It is also one
of the most effective techniques to produce pure nanocrystalline TiO_2 in powder form.

Anatase nano-TiO_2 powders were successfully prepared by a simple polymer gel
technique using polyvinylpyrrolidone (PVP) as the polymer. The products were systematically characterized using various techniques like TG/DTA, X-ray diffraction (XRD),
transmission electron microscope (TEM), and UV–visible spectroscopy. The TEM and
XRD reveal that the prepared powder had a pure anatase nano-TiO_2 structure. The nanosized materials were produced using a simple and cost effective polymer gel technique.

14.2   EXPERIMENTAL DETAILS

The hybrid matrix was prepared by the simple polymer gel technique using Tetraisopropylorthotitanante (TIOT) and PVP as precursors and ethanol as solvent. No catalysts were added. Measured amounts of TIOT and PVP were stirred magnetically with
ethanol at room temperature to avoid local inhomogeneities. The ratio of TIOT, PVP,
and ethanol in the solution was 1:1:10. The TiO_2-PVP gels, which were transparent
and viscous, were dried in an oven at 60°C for 24 hr, and then yellow solids were
obtained. These solids were grounded into powder. The prepared powder sample was
then calcined at 200°C and 350°C respectively. After calcinations, the sample was
further characterized. The characterizations were done with TEM (JEOL - JEM-2100
LaB6), X-ray diffractometer (Bruker AXS D8 Advance) and absorption with a spectrophotometer (Shimadzu-UVPC 2401).

14.3   DISCUSSION AND RESULTS

Figure 1 shows TG–DTA curves of the PVP-TiO_2 hybrid. A large weight loss is found
to start at temperatures around 384°C, indicative of the decomposition of PVP. The
endotherm at <100°C corresponds to the removal of adsorbed water. There are two
exotherms corresponding to temperatures 345 and 495°C. The former corresponds to
the burning of the organics and the latter could be due to decomposition of Titania gel
to anatase phase [4]. So the TGA results also seem to indicate that stable bonding exists between the organic and inorganic components.

Figure 2 shows the XRD spectra of the samples heat treated at 200 and 350°C. The
spectrum at 200°C directly shows that the samples are amorphous. The various peaks
found in the sample heat treated at 350°C reveals the crystallinity of the sample. The
peaks and the crystal planes were identified with the help of International Centre for
Diffraction Data (ICDD). Crystal planes 004, 101, 200, 211, 215, and 220, matched
well with that of the anatase phase of TiO_2. Using the Debye-Scherrer formula, the
crystallite size was calculated to be 4.8 nm from the XRD peak corresponding to 101
crystal plane. We also applied the Hall-Williamson correction factor [5] to the Scherrer formula to avoid any underestimation of crystal size by the line width broadening
due to lattice distortion. Eventually, an effective crystal size of 5.5 nm was obtained.
Figure 3 shows the TEM image revealing the high resolution morphology of the nanostructured particles containing anatase TiO_2 calcined at 350°C. It was observed that
Titania was almost exclusively composed of small nanocrystallites ranging in size
from 5 to 8 nm, and not much agglomeration even after heating at 350°C.

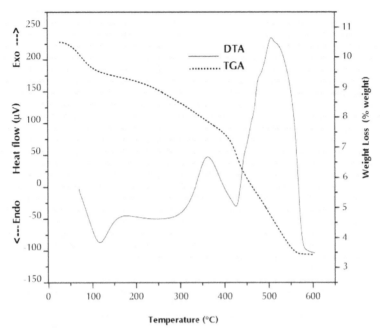

FIGURE 1   The TG/DTA spectrum of PVP-TiO₂ hybrid.

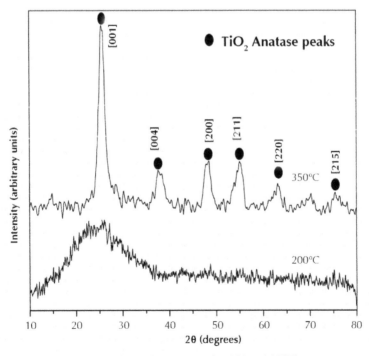

FIGURE 2   The XRD spectra of sample heat treated at 200 and 350°C.

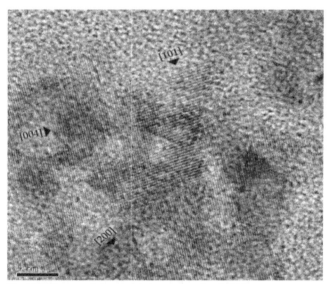

FIGURE 3   The TEM image of the sample heat treated at 350°C.

The optical absorption spectra of TiO_2 nanocrystallites in PVP matrix heat treated at 350°C is shown in Figure 4. The direct absorption band gap of the TiO_2 nanoparticles can be determined by fitting the absorption data to the Equation $\alpha h\nu = B(h\nu - E_g)^{1/2}$ (as shown in the inset of Figure 4) in which $h\nu$ is the photon energy, α is the absorption coefficient,

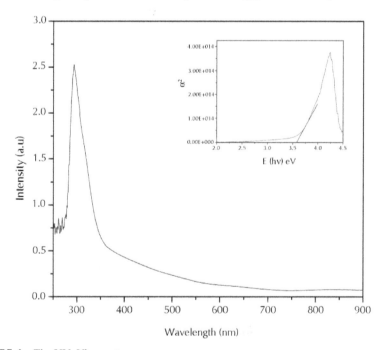

FIGURE 4   The UV–Vis spectra

E$_g$ is the absorption band gap and B is a constant relative to the material. The absorption coefficient can be obtained from the Equation $\alpha = 2.303 A/d$ where A is the absorbance and d is the thickness of the sample. The estimated band gap energy is 3.58 eV. This is large compared to bulk anatase TiO$_2$, a direct semiconductor, with band gap energy of 3.2 eV. Semiconductor nanocrystals are known to have an absorption edge, which is shifted with respect to the bulk material, toward shorter wavelengths [6].

14.4 CONCLUSION

Anatase TiO$_2$ nanocrystals were prepared by a simple polymer gel technique using PVP. The TEM and XRD results confirmed the formation of anatase TiO$_2$ nanocrystals at 350°C. The direct band gap of TiO$_2$ nanocrystals from the UV absorption spectrum was found to be 3.58 eV. It was concluded that the structure and morphology of the anatase TiO$_2$ nanoparticles formed by this simple polymer gel technique are similar to those of the product obtained by the tedious hydrothermal technique.

KEYWORDS

- Crystal plane
- Hydrothermal technique
- Polymer gel technique
- Tetraisopropylorthotitanante
- Transmission electron microscope

ACKNOWLEDGMENT

One of the authors (Prathibha Vasudevan) is thankful to KSCSTE, Trivandrum, Kerala, for award of Research fellowship. Authors are also thankful to UGC, Govt. of India for financial assistance through SAP-DRS program.

REFERENCES

1. Negishi, N., Matsuzawa, S., Takeuchi, K., and Pichat, P. *Chem. Mater.*, **19**, 13808–13814 (2007).
2. Haro-Poniatowski, E., De la Cruz Heredia, M., and Arroyo-Murillo, R. et al. *J. Mater. Res.*, **9**, 2102–2108 (1994).
3. Nuida, T., Kanai, N., Hashimoto, K., Watanabe, T., and Ohsaki, H. *Vacuum*, 74, 729–733 (2004).
4. Chen, L., Shen, H., Lu, Z., Feng, C., Chen, S., and Wang, Y. *Colloid Polym. Sci.*, **285**, 1515–1520 (2007).
5. Williamson, G. K. and Hall, W. H. *Acta metall.*, **1**, 22–31 (1953).
6. Lifshitz, E., Dag, I., Litvin, I., Hodes, G., Gorer, S., Reisfeld, R., Zelner, M., and Minti, H. *Chem. Phys. Lett.*, **288**, 188–196 (1998).

15 Thermoluminescence of Gd_2O_3:Er^{3+} Yb^{3+} Phosphors

Raunak Kumar Tamrakar, D. P. Bisen, and Nameeta Bramhe

CONTENTS

15.1 INTRODUCTION

Rare earth doped oxide particles have been widely studied for application to displays such as high definition (HD), projection televisions (PTVs), and flat panel displays (FPDs). Phosphor materials must have a narrow size distribution, non-agglomeration properties, and spherical morphology for good luminescent characteristics. The mean size of the particles is very important for high resolution and high efficiency [1].

The current interest in this field is focused on synthesizing phosphors materials using improved techniques and looking for their new applications. Out of various phosphors, Er^{3+}/Yb^{3+} doped Gd_2O_3 upconversion (UC) phosphors has been found to be highly efficient and has been successfully used in different applications [2-7]. The high quantum yield is due to low phonon energy of the Gd_2O_3 based host materials. Gd_2O_3 is excellent host matrix of UC luminescence because it has better chemical durability, thermal stability, and lower phonon energy [8, 9].

In recent years, numerous investigations have deal with upconverting phosphors compounds that produce emitted photons with a higher energy than the excitation photon energy *via* a two or more photon system. Such materials are composed typically of trivalent rare earth sensitizer (e.g., Yb^{3+} and Sm^{3+}) and activator (e.g., Er^{3+}, Ho^{3+}, Pr^{3+}, and Tm^{3+}) ions, acting in a multiphoton process [10].

Thermoluminescence (TL) is the emission of light observed during the heating of insulating or semiconductor materials, provided that they have been previously exposed to ionizing radiation [12-14].

The present chapter studies the combustion synthesis and TL studies of gadolinium phosphors doped with Erbium and Ytterbium. The TL glow curve and spectra of TL were measured, and TL response as a function of the absorbed with different UV exposer time. Then we also calculated kinetic parameters such as activation energy (E), order of kinetics (b), and frequency factor (s). The chapter deals mainly with the physical processes involved in the TL emission. It has been observed that TL intensity increases with increase of UV exposure time and the maximum TL intensity is found at 15 min exposer time for UV radiation. The value of activation energy belongs to 1.3321–1.4409 eV and frequency factor 1.52×10^{13} to 3.92×10^{14}, the order of kinetics is found to be first order.

15.2   EXPERIMENTAL

15.2.1   Sample Preparation

In this study gadolinium nitrate (99.99% Sigma Aldrich), erbium nitrate (99.99% Sigma Aldrich), and ytterbium nitrate (99.99% Sigma Aldrich), urea were used as starting raw material. To prepare Gd_2O_3: Er^{3+}, Yb^{3+}, these $Gd(NO_3)_3$, $Er(NO_3)_3$ and $Yb(NO_3)_3$ were mixed according to the stoichiometry equation in a beaker and then a suitable amount of urea was added to prepare the precursor solution and kept stirring for 30 min. Finally, this sample was transferred to crucible and fired in a furnace then water was evaporated quickly and soon a vigorous redox reaction occurred, the whole process went on for a few seconds at 600°C. Finally, Gd_2O_3: Er^{3+}, Yb^{3+} nanoparticles with different concentration were obtained.

15.2.2   TL Reader

The TL spectra recorded for Er^{3+} and Yb^{3+} doped Gd_2O_3 phosphors using a device TL reader (Model No. 1009-I) manufactured by Nucleonix System Private Limited Hydrabad. Samples were exposed by UV radiations from UVGL-58 handled UV lamp operating at 230 V–50 Hz (emitting 253 nm). The linear heating rate of the TL reader was set at 5°C/s. the kinetic parameter were obtained using only glow peak shape method (modified by Chen). Then, the kinetic parameters are calculated using the equation as suggested by Chen.

15.3   DISCUSSION AND RESULTS

15.3.1   TL Result

The TL is a very sensitive technique for the detection of traps or defects [15]. Figure 1 shows the TL glow curve of Gd_2O_3: Er^{3+}, Yb^{3+}. It is found that the TL glow curve is in quite increasing order for 5 and 10 min UV exposure and low intensity but for 15 min UV exposure we get the maximum intensity and the temperature decreases. The peak parameters (peak temperature, full widths, and shape factor) are shown in Table 1. The variation of shape factor (μ) implies that the first peaks is of first order kinetics. Since the shape factor is not unique for the intense peak one can speculate it as a consequence of overlapping of neighboring peaks.

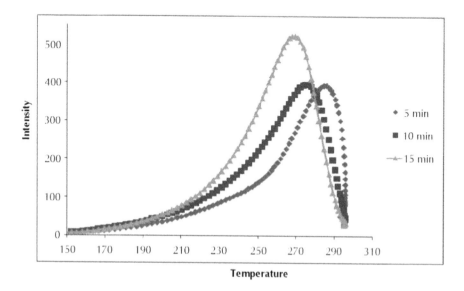

FIGURE 1   TL glow curve of Gd$_2$O$_3$: Er^{3+}, Yb^{3+} with different UV exposure time.

15.4   CONCLUSION

The TL property of Er^{3+} and Yb^{3+} doped Gd$_2$O$_3$ phosphors has been investigated for UV irradiation. The trapping parameters were calculated. The dependence of shape factor of TL peak exhibiting thermal quenching on the energy (W), characterizing the non-radiative process has been thoroughly studied. The shape factor of the first order kinetics is less than 0.42 whereas for a second order kinetics it may be vary from 0.52 to even 0.426 depending on the degree of thermal quenching. In this TL glow curve, μ ≈ 0.42 or less than first order kinetics may be caused by the presence of two or more traps with similar trap energies. We found here the value μ = 26 to 35 which show the first order kinetics. Also the value shows the trapping parameters and low fading because the curves shift to the lower temperature side. The value of activation energy belongs to 1.33–1.44 eV and frequency factor 1.52×10^{13} to 3.92×10^{14}, the order of kinetics is found to be first order.

TABLE 1 Kinetic Parameters of Gd_2O_3: Er^{3+}, Yb^3.

Exp. Time	T_1	T_m	T_2	$\tau = (T_m - T_1)$	$\delta = (T_2 - T_m)$	$\omega = (T_2 - T_1)$	$\mu = \delta/\omega$	Activation energy E	Frequency factor
5 min	528.21	557.53	568.3	29.32	10.77	40.09	0.26864	1.3321	1.52×10^{13}
10 min	519.99	547.97	562.36	27.98	14.39	42.37	0.33962	1.3598	4.69×10^{13}
15 min	516.28	542.21	556.22	25.93	14.01	39.94	0.35077	1.4409	$3zz.92 \times 10^{14}$

KEYWORDS

- **Doped phosphors**
- **Frequency factor**
- **Peak shape method**
- **Thermoluminescence**
- **Upconversion**

REFERENCES

1. Sievers, R. E., Milewski, P. D., Xu, C. Y., and Watkins, B. A. *Extended Abstracts of the Third International Conference on the Science and Technology of Display Phosphors*. Huntington Beach, California, p. 303 (1997).
2. Jang, H. S., Yang, H., Kim, S. W., Han, J. Y., Lee, S. G., and Jeon, D. Y. *Adv. Mater.*, **20**, 2696 (2008).
3. Lim, S. F., Riehn, R., Ryu, W. S., Khanarian, N., Tung, C. K., Tank, D., and Austin, R. H. *Nano Lett.*, **6**, 169 (2006).
4. Van de Rijke, F., Zijlmans, H., Li, S., Vail, T., Raap, A. K., Niedbala, R. S., and Tanke, H. *J. Nat. Biotechnol.*, **19**, 273 (2001).
5. Leblou, K., Perriat, P., and Tillement, O. *J. Nanosci. Nanotechnol.*, **05**, 1448 (2005).
6. Singh, S. K., Kumar, K., and Rai, S. B. *Sens. Actuators A*, **149**, 16 (2009).
7. Singh, S. K., Kumar, K., and Rai, S. B. *Appl. Phys. B*, **94**, 165 (2009).
8. Guo, H., Dong, N., Yin, M., et al. *J. Phys. Chem. B*, **108**, 19205–19209 (2004).
9. Guo, H., Li, Y., Wang, D., et al. *J. Alloy Compd.*, **376,** 23–27 (2004).
10. Auzel, F. E. *Proc. IEEE*, **61**, 758–786 (1973).
11. Guinier, A. *X-ray Diffraction*. Freeman, San Francisco (1963).
12. McKeever, S. W. S. *Thermoluminescence of solids*. Cambridge University Press, Cambridge, p. 390 (1985).
13. Martini, M. and Meinardi, F. *Thermally stimulated luminescence new perspectives in the study of defects in solids*. La Rivista del Nuovo Cimento, **20**-4(8), 1–71 (1997).
14. Chen, R. and McKeever, S. W. S. *Theory of Thermoluminescence and RelatedPhenomenon*, World Scientific, Singapore, New Jersey, London, Hong Kong, p. 559, (1997).
15. Sharma, Ravi, Chandra, B. P., and Bisen, D. P. *Chalcogenide Letters*, **6**(6) 251–255 (June, 2009).

16 Thermoluminescence and Kinetics of Copper Doped CdS Nanoparticle

Raunak Kumar Tamrakar, D. P. Bisen, and Nameeta Brahme

CONTENTS

16.1 INTRODUCTION

Materials with nanoscopic dimensions such as quantum dots, nanowires, nanorods, and nanotubes have attracted a great deal of attention during the last four decades [1, 2]. Due to their applications in solar cells [3], catalysis [4], light emitting devices [5, 6], resonant tunneling devices [7], and lasers [8]. Blue shift in the optical absorption spectra, size dependent luminescence, enhanced oscillator strength, and nonlinear optical effect is some of the interesting properties exhibited by these nanocrystals. It was observed that the particle size decreases with decreasing synthesis temperature [9, 10]. The doping of transition metal ion such as Mn, Cu, Co, and so on opens up possibilities of forming new class of material and new properties of the materials are expected. The transition metal doped nanoparticles show different optical properties corresponding to their host counterparts. These nanoparticles have found tremendous application in optical light emitting diodes [11-14].

Thermoluminescence (TL) is a common and widespread phenomenon, which can be concisely described as emission of light at characteristic temperatures from samples that had been exposed to electromagnetic or particle radiation prior to their warming up in the dark. The most widespread applications of TL phenomenon is the radiation dosimetery in health physics, biological sciences, and radiation protection. Besides this TL is a very sensitive technique for the detection of traps or defects [15-16]. The TL of ZnS:Mn nanoparticles is also reported [17].

The present chapter reports that the thermally stimulated luminescence and kinetics of Cu doped cadmium sulfide (CdS) nanoparticle with its trap parameters. It has been observed that TL intensity increase with increase of UV exposure time and the maximum TL intensity is found at 15 min exposer time for UV radiation. The absorption spectra are observed and reported. The optical absorption spectra of CdS in the range of 400–700 nm and the band gap energy are found between 2.5 and 2.62 eV. We also calculate the kinetic parameters such as activation energy (E), order of kinetics (b) and frequency factor (s) for Cu doped CdS nanoparticles. The activation energy found between 1.14 and 1.72 eV and frequency factor of the order of 9×10^8–3.2×10^{10}, the order of kinetics is found to be first order.

16.2   EXPERIMENTAL DETAILS

16.2.1   Synthesis of CdS:Cu

Nanoparticles of copper doped CdS are prepared by wet chemical route method. For synthesis, 10^{-2} M aqueous solution dilutes solutions of copper acetate $Cu(CH_3COO)_2$ and sodium sulfides (Na_2S) were mixed in presence of (capping agent) thioglycerol ($C_3H_8O_2S$).

The aqueous solution of $C_3H_8O_2S$ was added drop wise in the solution of cadmium acetate with the help of burette at the rate of 1 ml/min while stirring the solution continuous and Na_2S was mixed drop by drop in an ice bath with constant stirring into the solution. Subsequently, a yellow color solution was obtained. This solution was kept 24 hr till precipitate settles down in the bottom of the flask. This precipitate is removed and washed several times with double distilled water. The unreacted $C_3H_8O_2S$ and Na_2S are removed by washing the solution several times. This washed solution was centrifuged then finally precipitate was spread over a glass substrate and air dried at room temperature. Similar method is used to prepare the samples of CdS:Cu for 10^{-1} and 1 M concentrations of the capping agents.

Absorption spectra of the samples prepared with various concentrations of capping agent and at different temperature were studied. The absorption spectra of the samples were recorded with the help of Shimadzu UV/VIS-1700 spectrophotometer.

16.2.2   TL Reader

For recording TL, samples were exposed to UV radiations from UVGL-58 handled UV lamp operating at 230V-50 Hz (emitting 253 nm) for 10 min. The TL glow curve were recorded on a Harshaw TLD Reader (Model 3500) fitted with 931B photomultiplier tube (PMT) by taking 5 mg of sample each time.

16.3   DISCUSSION AND RESULTS

16.3.1   Absorption Spectra

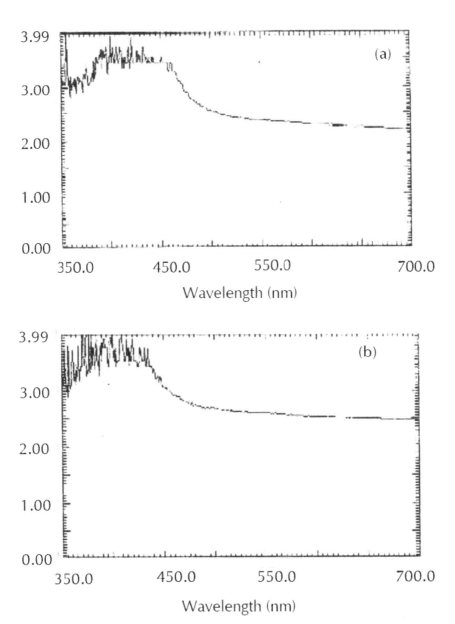

FIGURE 1   (a) Optical absorption curve of Pure CdS and (b) Optical absorption curve of CdS:Cu.

Figure 1 (a) and (b) shows the optical absorption spectra of pure and Cu doped CdS in the range of 400–700 nm. Optical absorption edge was obtained at 500 nm for pure CdS and at 474 nm for Cu doped CdS. The band gap energy of the samples corresponding to the absorption edge is found pure CdS is 2.5 eV and for 2.62 eV for Cu doped CdS.

16.3.2   TL Result

The TL is well known phenomenon that caused by the thermally assisted release of the irradiation induced electrons from the traps of the material. The TL curve can provide valuable information of the intrinsic defect of materials and the energetic ray to which the material was subjected. Therefore, it is widely applied in the field of defect studying, dating in Archeology, irradiation detection, and so on [18].

The TL glow curve of Cu doped CdS nanoparticles shows with different concentration in Figure 2. The TL glow curve for CdS:Cu for a UV exposure of different time at a 10 degree/s. The order of kinetics and the activation energy with frequency factor of this glow curve was found using Chen's empirical formula [19].

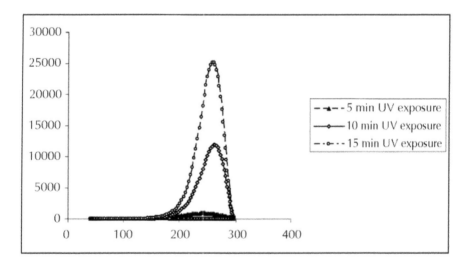

FIGURE 2   TL glow curve of Cu doped CdS nanoparticles.

16.4   CONCLUSION

The TL property of CdS doped with Cu nanoparticles has been investigated for UV irradiation. The trapping parameters were calculated. The generation and analysis of TL spectra of luminescent materials carried out in this study show that the shape, intensity, and position of a glow curve can depend not only on the intrinsic parameters of

TABLE 1 Kinetic parameters with different percentage of Cu for different UV exposure time.

UV exp time	T_1	T_m	T_2	$\tau = (T_m - T_1)$	$\delta = (T_2 - T_\mu)$	$\omega = (T_2 - T_1)$	$\mu = \delta/\omega$	Activation Energy (eV)	Frequency factor s
05	474.68	520	552.68	45.32	32.68	78	0.418974	1.144269	$.9 \times 10^8$
10	508.21	539	561.96	30.79	22.96	53.75	0.427163	1.716546	1.8×10^{10}
15	508.97	540	561.97	31.03	21.97	53	0.414528	1.797256	3.2×10^{10}

the relevant trap but also on the presence of other traps, on the presence of preionized luminescence centres, and on the level of excitation used to create the TL spectrum. Glow curves characterized by the value of order of kinetics b. In this TL glow curve, $\mu \approx 0.42$ or less than first order kinetics may be caused by the presence of two or more traps with similar trap energies. Also the value shows the trapping parameters and low fading because the curves shift to the lower temperature side. The trapping parameters were calculated. The phosphors CdS:Cu is found to have first order kinetics in TL emission suggesting retrapping of charges. The activation energy E is found in between 1.14 and 1.79 and the frequency factor is found in between 9×10^8–3.2×10^{10}

KEYWORDS

- **Activation energy**
- **Chen's empirical formula**
- **Nanoparticles**
- **Photomultiplier tube**
- **Thermoluminescence**

REFERENCES

1. Peng, W. Q., Cong, G. W., Qu, S. C., and Wang, Z. G. *Opt. Mater.*, **29**, 313 (2006).
2. Bouvy, C., Piret, F., Marine, W., and Su, B. L. *Chem. Phys. Lett.*, **433**, 350 (2007).
3. O'Reagan, B and Gratzel, M. *Nature*, **353** 737 (1991).
4. Panjonk, G. M. *Appl. Catal.*, **72**, 217 (1991).
5. Lazarouk, S. et al. *Appl. Phys. Lett.*, **68**, 1646 (1996).
6. Bhargava, R. N., Gallagher, D., Hong, X., and Nurmikko, A. *Phys. Rev. Lett.*, **72**, 416 (1994).
7. Nicolian, E. H. and Tsu, R. *J. Appl. Phys.*, **74**, 4020 (1993).
8. Schmitt-Rink, S., Miller, D. A. B., and Chemla, D. S. *Phys. Rev. B*, **35**, 8113 (1987).
9. Nogriya, V., Dongre, J. K., Ramrakhiani, M., and Chandra, B. P. *Chalcogenide Letters*, **5**(12), 365–373 (December, 2008).
10. Bisen, D. P., Sharma, R., Brahme, N., and Tamrakar, Raunak. *Chalcogenide Letters*, **6**(9), 427–431 (September, 2009).
11. Dong, L. S., FU, X. F., Wang, M. W., and Liu, C. H. *J. Lumin.*, **87–89**, 538 (2000).
12. Francois, N., Ginzberg, B., and Bilmes, S. A. *J. Sol Gel Scitechn.*, **13**, 341 (1998).
13. Premachandran, R., Banerjee, S., John, V. T., Mcpherson, G. L., Akkara, J. A., and Kaplan, D. L. *Chem. Mater.* **9**, 1342 (1997).
14. Gao, M. Y., Zhang, X., Yang, B., Li, F. and Shen, J. C. *Thin Solid Films.* **284-285**, 242 (1996).
15. Chen, R. and Kirsh, Y. *Analysis of Thermally Stimulated Processes*. Pergamon Press, Oxford (1981).
16. Vij, D. R. (Ed.). *Thermoluminescent Materials*. PTR Prentice Hall, New Jersey (1993).
17. Sharma, R., Chandra, B. P., and Bisen, D. P. *Chalcogenide Letters*, **6**(6), 251–255 (June, 2009).
18. Nagbhushana, H. PhD. thesis, Bangalore University, Bangalore (2003).
19. Chen, R. and Krish, Y. *Analysis of Thermally Stimulated processes*. Pergamon Press, New York (1981).

17 Poly(Vinyl Alcohol)/ Poly(Methacrylate-Co-N-Isopropylacrylamide) Composite Thermoresponsive Microgels for Drug Delivery

S. V. Ghugare, B. Cerroni, E. Chiessi, R. Fink, and G. Paradossi

CONTENTS

17.1 INTRODUCTION

Controlled drug delivery systems can be strictly modulated by stimuli in order to treat diseases in which sustained drug release is undesirable. [1, 2] Among the many kinds of stimuli-sensitive delivery systems, temperature-sensitive drug delivery systems offer great potential over their counterparts due to their versatility in design, tunability of phase transition temperatures, passive targeting ability and *in situ* phase transitions. [3] Thus, thermosensitive drug delivery systems can overcome many of the hurdles of

conventional drug delivery systems in order to increase drug efficacies, drug targeting, and decrease drug toxicities. In an effort to further control existing temperature-
responsive systems, current innovative applications have combined temperature with
other stimuli such as pH and light. The result has been the development of highly
sophisticated systems which demonstrate exquisite control over drug release and represent huge advances in biomedical research [4].

A controlled drug delivery system is designed to maintain the drug release to the
target site with a minimal adverse effect. [5, 6] Although many new drugs have been
discovered, the means to improve the efficiency of these new drugs *via* proper delivery
remains a challenging issue. Traditionally, oral or parenteral administration of drugs
leads to uncontrolled fluctuations in drug concentration in the bloodstream which can
results in undesirable adverse effect. Therefore, the development of effective drug car-
riers could be as important as the discovery of new drugs.

Various thermoresponsive microgels have been developed for drug delivery based
on poly(*N*-isopropyl acrylamide) (PNIPAAm), since its unique volume phase transition temperature (VPTT) in water is around 32°C, which is close to human body
temperature. [7, 8] These microgels have an ability to swell and shrink significantly
in aqueous environment at temperature below and above the volume phase transition
temperature, respectively.

Recently, we have developed thermoresponsive microgels based on poly(vinyl alcohol) (PVA) using a water-in-water microemulsion technique which has proved to
be handy and safe approach to obtain the microgels based on polymer incompatibility
and incorporated *N*-isopropyl acrylamide (NIPAAm) while maintaining other chemical and structural properties. [9-11]

The direct characterization of the shrinking behavior of the microgel induced by
temperature changes in aqueous solution is important to understand their colloidal
features. Differential scanning calorimetry (DSC) is a convenient technique for measuring the temperature-dependent change in enthalpy of the microgel. However, this
technique cannot provide any morphological or spectroscopic information. There are
several imaging techniques for colloidal particles dispersed in aqueous media. Optical microscopy (OM) is useful for sizing micron-sized particles in solution. However,
because of the wavelength of visible light and other constraints, the effective OM
resolution limit is approximately 0.5 μm. Moreover, in the particular case of microgels, it is very difficult to observe even micrometer-sized particles directly because of
the limited difference in refractive index between the swollen particles and the continuous phase. Recently, atomic force microscopy (AFM) was used to image hydrated
microgel particles adsorbed on mica in aqueous solutions, thereby avoiding the gross
structural changes that can occur upon drying. However, no spectroscopic information
can be obtained with this technique [12].

Within light microscopy methodologies, confocal laser scanning microscopy
(CLSM) is also a powerful technique for observing microgel particles in the wet state.
However, this method also suffers from relatively low resolution, and the microgels
must be fluorescently labelled which may affect their shrinking properties. Scanning
transmission X-ray microscopy (STXM) combines excellent compositional sensitivity as near-edge X-ray absorption fine structure (NEXAFS) [13] spectroscopy can be

simultaneously performed with high spatial resolution and has been recently used to study a number of polymer systems. [14] The socalled "water window" (E 284-543 eV) corresponding to photon energies between the K-shell absorption edge of carbon and oxygen allows STXM characterization of fully hydrated organic samples such as biofilms [15, 16] and synthetic polymers [17-21].

The PVA is known for its hydrophilic character and is often used to maintain high hydration levels in devices when closely interacting with tissues. The NIPAAm residues at room temperature are hydrophilic as well. The volume phase transition occurring at physiological temperature is due to the desolvation driven by the increase of hydrophobicity of NIPAAm in these conditions. Therefore, it is important to study the cell interaction of these microgels.

We describe the study of the thermoresponsiveness of PVA/Poly(MA-co-N-isopropyl acrylamide) microgels in this chapter. In particular the direct visualization of the shrinking behavior of microgel around its VPTT by CLSM and STXM. Furthermore, this chapter addresses the drug loading, release kinetics, cytotoxicity of doxorubicin (DOX) loaded microgels, and cell interaction with thermoresponsive PVA/Poly(MA-co-N-isopropylacrylamide) microgels were studied towards the human adenocarcinoma tumor cell line, HT29.

17.2 EXPERIMENTAL

17.2.1 Materials

The NIPAAm, purchased from Aldrich, was recrystallized in n-hexane prior to use. All other chemicals were used as received. Fetal bovine serum (FBS), phosphate buffered saline (PBS), pH 7 4, succinic anhydride, Phalloidin-FITC, thiazolyl blue tetrazolium bromide (MTT), and DOX PVA with a number and weight average molecular weights of 30,000 ± 5,000 and 70,000 ± 10,000 g/mol, respectively, dextranT40 with average molecular weight of 35,000–40,000 g/mol were purchased from Sigma-Aldrich. 4-(N, N-Dimethylamino) pyridine (DMAP), glycidylmethacrylate (GMA), and fluorescein isothiocynate isomer 1 (FITC) were Fluka products. Photoinitiator 2-hydroxy-1-[4-(hydroxyethoxy)phenol]-2-methyl 1-propanone (Irgacure 2959) was purchased from Ciba. Mc Coy's 5A medium, L-glutamine 200 mM and penicillin/streptomycin solution, at concentration of 10,000 U/ml and 10 mg/ml, respectively, were purchased from HyClone. Human adenocarcinoma cells, HT-29 cell line was purchased from "Istituto zooprofilattico sperimentale della Lombardia e dell'Emilia Romagna" Brescia, Italy. Trypsin 10X solution was purchased from Lonza (Basel, Switzerland). Dimethyl sulfoxide (DMSO), inorganic acids, and bases were RPE grade products supplied by Carlo Erba (Italy). Water was Milli-Q purity grade (18.2 MΩ.cm) produced with a deionization apparatus (Pure Lab) from USF, Elga. Dialysis membranes (cut off 12,000 g/mol) were purchased from Medicell International Ltd and prepared according to standard procedure.

17.2.2 Methods

Synthesis of Methacylate PVA (Pva-Ma)

The PVA was grafted with glycidyl methacrylate according to the procedure described elsewhere. [9, 22] Typically, 10 g of PVA was dissolved in 250 ml of DMSO at 70°C.

After complete dissolution of PVA, the flask was cooled to room temperature, and then 5 g of DMAP was added under N_2 flux. After 1 hr, a known amount of GMA was added in the proper molar ratio with respect to moles polymer repeating unit. The reaction was carried out at room temperature in darkness under stirring and stopped after 48 hr by neutralizing the solution with an equal molar amount of HCl with respect to DMAP. The DMSO was replaced with water by exhaustive dialysis. The sample was stored as a freeze-dried powder. After methacryloyl grafting, the degree of substitution (DS), was determined by H NMR spectroscopy. For microgels preparation, a sample with DS 5%, hereafter labeled PM5 was used throughout this study.

Preparation of PVA Based Microgels

Microgels based on PM5 were prepared, adapting a procedure originally introduced by Franssen and Hennink, [23] which employs a water-in-water emulsion technique based on polymer-polymer immiscibility in aqueous solutions. In a typical experiment, an aqueous dispersion containing dextran T40 at a concentration of 16% (w/v), PM5 at 2% (w/v), 1.3% NIPPAm, and the UV photoinitiator Irgacure 2959 at 0.3% (w/v) was vigorously stirred at 16,000 rpm by an UltraTurrax emulsifier. After emulsification, PM5 in the dispersed aqueous phase was cross-linked by photopolymerization using a 365 nm light source at an intensity of 7 m W/cm^2 for 5 min. The NIPAAm monomers, present in both aqueous phases, were cross-linked in the PM5 containing phase because of the presence of vinyl moiety grafted to PVA chains. The cross-linked PVA/Poly(methylmethacrylate-co-*N*-isopropyl acrylamide) microgels were purified by repeated steps of centrifugation and resuspension in Milli-Q water. Microgels will be hereafter labeled as PMN-II.

Confocal Laser Scanning Microscopy (CLSM)

Freeze-dried PMN-II microgels were first immersed in Milli-Q water at room temperature to reach their equilibrium state. The swollen microgels were labelled with FITC by coupling the dye to hydroxyl group of microgels. [24] Fluorescent dye at a concentration of 10 μm was added in the dispersion. The excess dye was removed by repeated washings with Milli-Q water, and CLSM observations were performed using a confocal laser scanning microscope (Nikon PCM 2000, Nikon Instruments, Japan) with a Plan Apo 60XA/1.4 oil immersion objective. The 488 nm line of 100 mW argon ion laser (Spectra-physics Lasers, USA) was used for sample fluorescence excitation of microgels. The VPTT was examined at room temperature and at 40°C. These measurements were performed using a Bioptechs – FCS2 and FCS 3 closed chamber stage with a built-in temperature control unit.

Scanning Transmission X-ray Microscopy (STXM)

For STXM measurements we used the socalled "wet cells" where approximately 1 μl of well homogenized microgel water suspension was sandwiched between two 100 nm thick Si_3N_4 membranes (Silson Ltd, UK), which were then sealed with silicone high vacuum grease to maintain the water environment during the experiment. The microgels were imaged in transmission mode in helium atmosphere using the PolLux STXM microscope at the Swiss Light Source (SLS), using the synchrotron radiation of the electron storage ring of the Paul Scherrer Institute (Villigen, Switzerland). The transmitted photon flux was measured using a photomultiplier tube (Hamamatsu 647P). The PolLux STXM uses linearly polarized X-rays from a bending magnet in the photon energy range between 260 and 1100 eV and it routinely provides a spatial resolution better than 40 nm [25].

Images were recorded at selected energies through the O 1s region (510–560 eV). The data for the beam induced changes of the microgels were obtained during a normal "line-by-line" imaging experiment with a dwell time of 1 ms (data integration time per one image pixel). Carbon K-edge NEXAFS spectra were collected in line mode, that is, the transmitted signal was recorded while a line trajectory was scanned across the center of microgel particle at each value of the photon energy through the spectrum. Data processing was carried out using the aXis2000 software [26]. Radial transmittance profile processing was performed with homemade software.

Preparation of Negatively Charged Microgels

Details on the preparation of negatively charged PM-I type microgels has been already reported [9, 22] and was also applied to PMN-II. Typically, 1 g of PMN-II microgels and 0.5 g DMAP were suspended in 25 ml of anhydrous DMSO and stirred under N_2 flux. After 1 hr, succinic anhydride (0.25 g) was added, keeping the suspension at 40°C for 48 hr under nitrogen atmosphere. The reaction was stopped by neutralization and the suspension was dialyzed against Milli-Q water. The degree of O-succinoylation of the microgels, determined by potentiometric titration, was 10 ± 2 mol% with respect to PVA repeating units.

DOX Loading

In typical DOX loading experiment, 10 mg of freeze-dried O-succinoylated microgels was suspended in 3 ml of aqueous DOX at a concentration of $1.70 \cdot \times 10^{-4}$ M and left for 12 hr under gentle stirring. Then microgels were centrifuged and the absorbance of supernatant measured at 485 nm with a Jasco V-630 double beam UV-vis. Determination of the DOX payload was based on a molar extinction coefficient value of $\varepsilon_{485} = 11,500$ $M^{-1}cm^{-1}$.

DOX Release Kinetics

Typically, 10 mg of freeze dried DOX loaded PMN-II microgels immersed in 3 ml of PBS solution at a controlled temperature of 20°C and 37°C. At regular times the supernatant from each sample was removed after centrifugation and replaced with the same amount of fresh PBS solution. The DOX release was monitored by measuring the time dependence of the absorbance at 485 nm. Release behavior was analyzed in terms of cumulative ratio (%) of released DOX.

Cell Culturing

The HT-29 human colon adenocarcinoma cells were maintained at 37°C in a humidified atmosphere of 5% CO_2 (HERA cell 150i CO_2 incubator) in cell culture medium Mc Coy's 5A supplemented with 10% FBS, 2 mM L-glutamine, 100 U/ml penicillin and 100 µg/ml streptomycin (complete medium).

MTT (methylthiazole tetrazolium) Cytotoxicity Assay

The HT-29 cells in exponential growth phase were washed three times with PBS, trypsinized, re-suspended in serum free medium and seeded in 24-multi well plates. Cells were incubated for 12 hrs to permit the adhesion and the formation of a partial monolayer. They were than exposed to various quantities and types of microgels in complete medium and they were incubated at 37°C and 5% CO_2 for different times.

The MTT, dissolved in PBS solution (5 mg/mL), was added at fixed times in the wells up to a concentration of 10% of the total culture volume to form formazan purple crystals as reaction product with cellular mitochondrial metabolites [27]. After 4 hr, the solution over the crystals was removed and replaced with DMSO, a good solvent for the water-insoluble formazan. Plates were further incubated for 5 min at room temperature and the absorbance at 570 nm was recorded by UV spectroscopy (Jasco V-630, Japan).

For both cytotoxicity and proliferation studies, microgels were sterilized by exposing them at UV lamp for 20 min.

Cell Staining

Cells were seeded on Thermanox™ Coverslip Sterile (Nunc) in 24-multi well and incubated at 37°C in 95% air/5% CO_2 environment overnight. Then, microgels were incubated with cells for different times.

Cells were stained using phalloidin-FITC which provides a green fluorescent labeling of the cytoskeleton due to the specific binding with F-actin. Phalloidin-FITC was solubilized in water to form a stock solution with a concentration of 0.1 mg/ml. Cells, grown on sterile coverslips, were rinsed twice with PBS buffer solution and then fixed for 10 min in 3.7% formaldehyde in PBS. After washing with PBS, cells were permeabilized with 0.1% Triton-X 100 in PBS for 7 min. Cells, washed with PBS, were stained using a 50 μg/ml fluorescent phalloidin-FITC solution in PBS. After 40 min at room temperature, cells were rinsed with PBS twice and samples were observed with a confocal laser scanning microscope Nikon Eclipse Ti-E with a Plan Fluor 100X 1.30 oil immersion objective.

17.3   DISCUSSION AND RESULTS

Recently, group developed thermoresponsive microgels based on PVA copolymerized with NIPAAm. The microgels were obtained with a spherical shape having an average diameter of around 2 μm with relatively narrow size distribution. [9] It has been also investigated in details of the VPTT and the diffusion process of confined water as well as the segmental chain motion of the polymer network around its volume phase transition temperature. Furthermore, we have investigated the permeation of FITC-dextran molecule around the microgel volume phase transition temperature. [10] This temperature dependent response allows the microgel to be used, at least potentially, as a platform for thermally activated drug release in injectable drug delivery system. Indeed, the *in vitro* release of DOX, a well studied anti tumor drug, was enhanced when microgels loaded by DOX were heated from room temperature to physiological temperature.

The polymeric moiety of the microgel network is made by PVA linked by poly(methacrylate-co-NIPAAm) copolymers as described by Scheme 1.

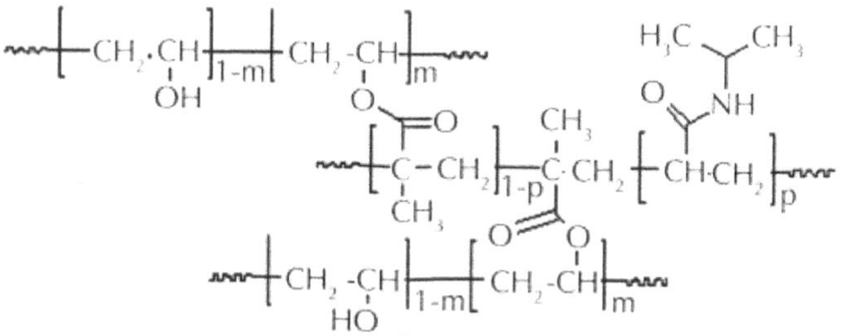

SCHEME 1   Chemical structure of PVA based network.

17.3.1   Shrinking Behavior of PMN-II Microgel Particles

Ade et. al. [28, 29] has investigated the first direct imaging of electrolyte-induced deswelling behavior of pH-responsive microgels in aqueous media using STXM.

Here the similar approach has been employed for the thermoresponsive PMN-II microgels using CLSM and STXM.

In the Figure 1, CLSM image shows the direct shrinking behavior of single thermoresponsive PMN-II microgel. This study allow to assess that above the VPTT the microgels shrink and the area reduction ratio of PMN-II microgel is around 22%. This result is in accordance with the STXM results.

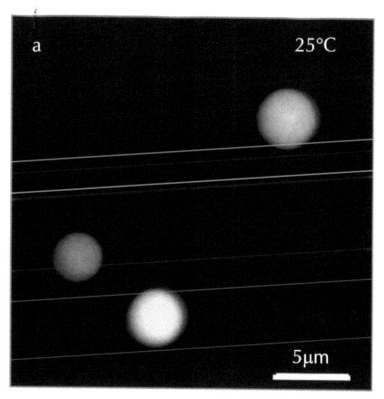

FIGURE 1   *(Continued)*

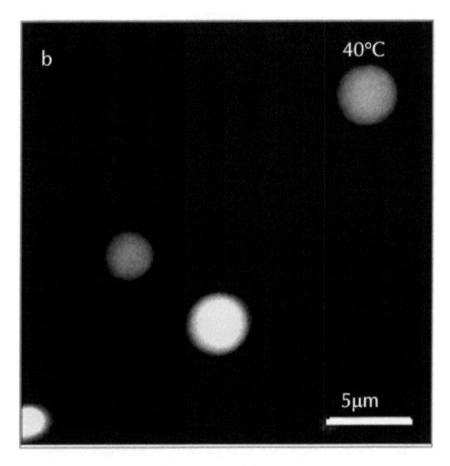

FIGURE 1   The CLSM image of PMN-II microgels at 25 and 40°C.

The STXM transmission images of microgels (PMN-II) are shown in Figure 2 below and above the VPTT. The images were recorded at 520 eV of a same scan region. The images were recorded following the same microgel particles at different temperature.

The change of the radial transmittance profile for one of the microgel particle as a function of temperature is presented in Figure 3 (a) and it revealed that the microgel particle shrink at 45°C by 15% of diameter.

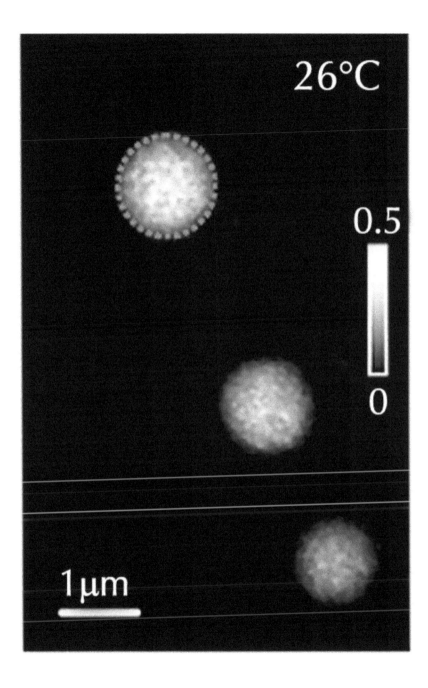

FIGURE 2 *(Continued)*

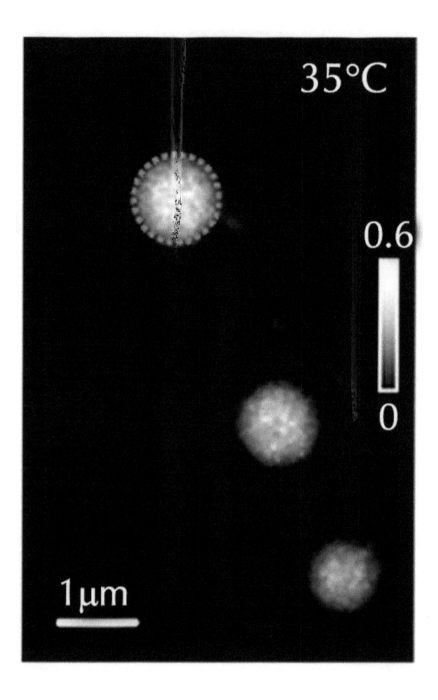

FIGURE 2 *(Continued)*

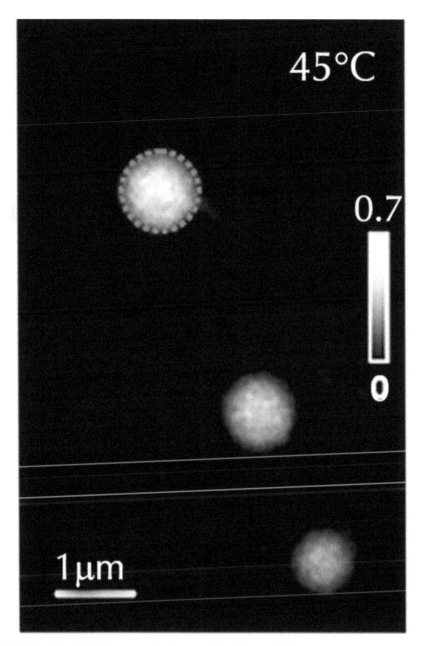

FIGURE 2   The STXM transmission image at 26, 35, and 45°C.

We have not showed the radial transmittance profile for the other microgel particles presented in the Figure 2, however those particles also shrink at 45°C by 15% of diameter.

(a)

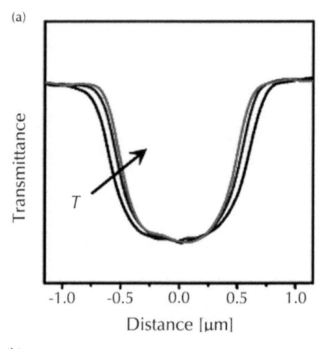

(b)

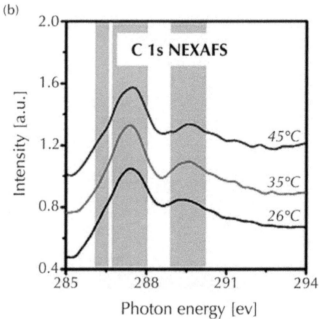

FIGURE 3 Radial transmittance profile of STXM images at 26°C (black), 35°C (blue), and 45°C (red) (a) and C 1s NEXAFS at 26°C (black), 35°C (red), and 45°C (blue) (b) of thermoresponsive microgels.

In order to rule out any chemical transformation of the microgel around the VPTT the chemical composition of the microgels above and below the VPTT carbon K-edge NEXAFS spectroscopy was applied. Absorption spectrum of the microgel particles below and above VPTT is shown in Figure 3 (b). Besides very minor intensity variations which are beyond the sensitivity of the technique, the T-dependent NEXAFS spectra confirm that the microgel chemical structure remained unchanged above the VPTT.

17.3.2 DOX Drug Loading and Release

In order to investigate the influence of the temperature on the release properties of PMN-II microgels, the networks were loaded with DOX, one of the most common anticancer drugs (Figure 4).

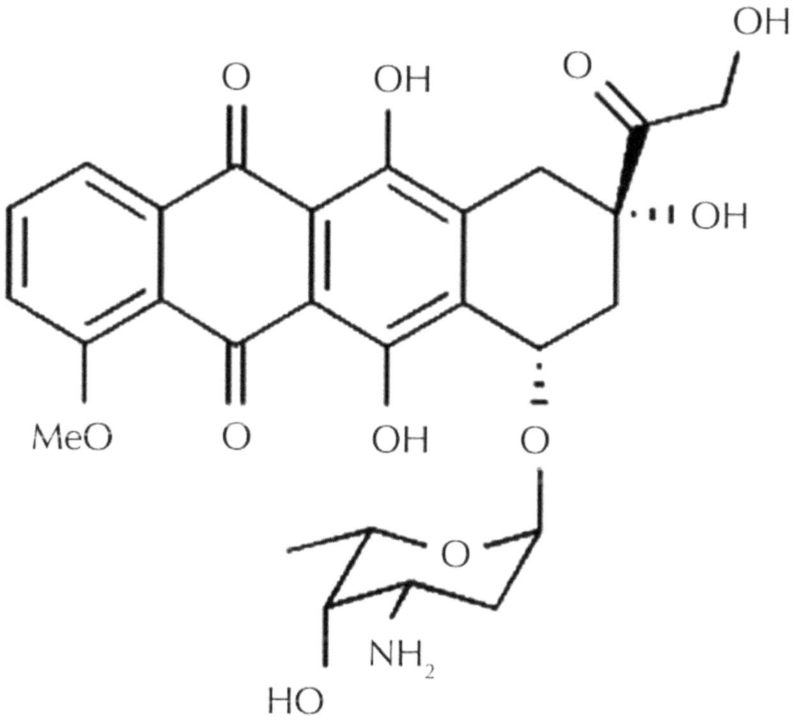

FIGURE 4 Chemical structure of DOX.

To optimize the cargo payload, we adopted the same strategy used for PMN-I microgels, [9, 22] conjugating succinoyl groups to the hydroxyl moiety of the microgel (Scheme 2). In this way, PMN-II microgels were functionalized with a 10% of O-succinoylation with respect to PVA repeating units. This substitution provided a surface negative charge density, favoring the adsorption of DOX molecules bearing positively charged amine groups at physiological pH. The loading was accomplished by suspending the microgels in an aqueous solution of DOX.

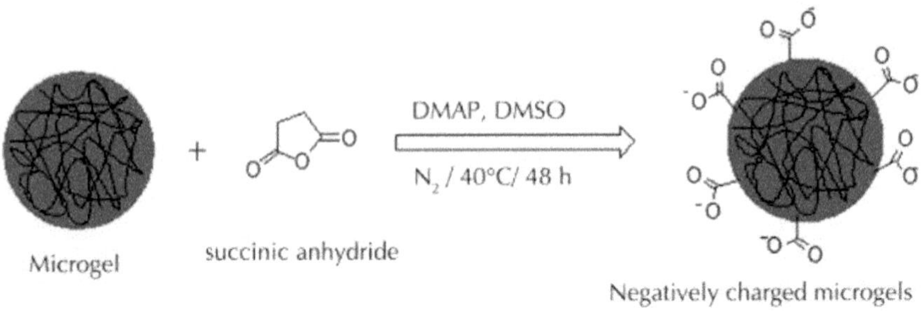

SCHEME 2   Reaction scheme of O-succinoylation.

The loading efficiency of DOX was 92 ± 3 (M%). Loaded microgels were stored as freeze-dried powder.

The release kinetics of succinoylated PMN-II microgels were studied in PBS. The cumulative drug release as a function of time has been plotted in Figure 5. The data were fitted by a pure Fickian diffusion equation:

$$\frac{M_t}{M_\infty} = kt^{0.5}$$

The good agreement of this model with the experimental behavior indicates that the DOX release is not influenced in the explored time window by potential matrix degradation.

At equilibrium the final amount of released drug at room temperature was 45 M%, at 37°C where the cumulative release at equilibrium is 65 M%. At this temperature PMN-II microgels completed the volume phase transition and the larger release of DOX exhibited by PMN-II corresponded to a more pronounced shrinking effect occurring in this system (Figure 6).

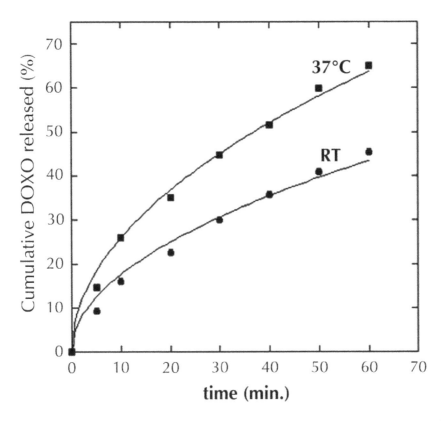

FIGURE 5   The DOX release kinetics of (●), (■),and PMN-II in PBS at room temperature and at 37°C respectively.

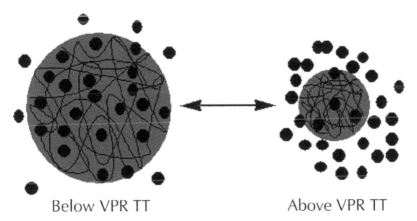

Below VPR TT   Above VPR TT

FIGURE 6   Schematic representation of DOX release from DOX loaded microgel below and above VPTT.

The more efficient drug release exhibited by PMN-II at higher temperature was the result of the increased hydrophobicity of the microgel networks at 37°C. This attributed to making the interaction with charged DOX molecules less favorable, and increase of the specific area due to the volume phase transition occurring in microgels when suspensions are brought from room to physiological temperature.

17.3.3   Cytocompatibility of PMN-II Microgels

The MTT assay experiments were carried out to test the cytocompatibility of PMN II microgels on colon adenocarcinoma cell, HT-29. 50,000 HT-29 cells were incubated with different amounts of sterile PMN-II. Data shown are averages of three independent experiments (Figure 7).

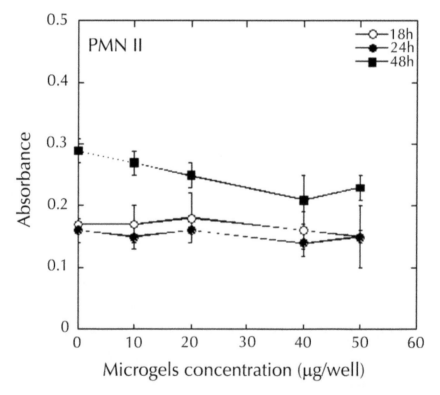

FIGURE 7   Cytocompatibility of HT-29 human adenocarcinoma cell line incubated with different amounts of PMN-II for 18 hr (O), 24 hr (●), and 48 hr (■), respectively.

At all concentrations (from 10 to 50 μg) PMN-II microgels showed cytocompat-ibility after an incubation of 18 and 24 hr with HT-29. When HT-29 cells are incubated for 48 hr, there is a trend towards a decrease in HT-29 viability of about 20% when increasing microgels concentrations.

Further MTT assays were also performed to assess viability of HT-29 when they were incubated with PMN-II microgel loaded with anticancer chemotherapy drug, DOX (Figure 7). 40,000 cells were seeded in 24 multiwells and then viability was measured.

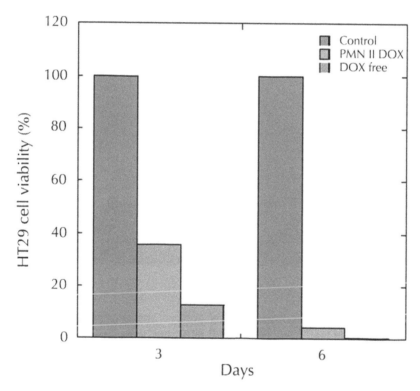

FIGURE 8 Cytocompatibility test (MTT of PMN-II on HT- 29 cells).

Figure 8 shows the comparison between the cells growth of untreated HT-29 cells to HT-29 cells incubated with 50 μg of PMN-II loaded with DOX. The amount of free DOX added to cells was the same of the DOX loaded on 50 μg of PMN II particles.

As expected, HT-29 cells treated with PMN-II loaded with DOX show a low vi-ability after 3 days that become close to 0 after 6 days. Free DOX kills cells in a more effective way then PMN-II, this suggests that these particles could represent a vehicle to release drug (DOX) in a controlled way.

17.3.4   Cell Interaction

In order to investigate the biointerface behavior of PMN-II microgels, we addressed preliminary study of interaction with HT-29 cell line. As shown in Figure 9 the interaction between PMN-II loaded with DOX and HT-29 cell line was investigated by CLSM. Cells were incubated with 50 µg on PMN-II loaded with DOX for different times (1, 3, and 6 days) and the uptake of PMN-II loaded with DOX by cells was observed (Figure 8). Increasing the incubation time between microgels and cells, HT-29 changed their morphology and decreased in quantity because of the presence of DOX. It was also possible to see cell uptake of particles PMN II by HT-29 cells.

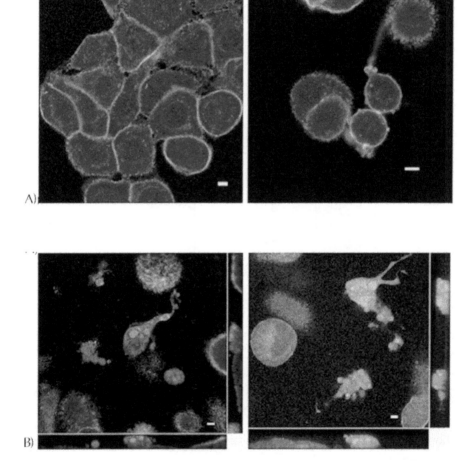

FIGURE 9   *(Continued)*

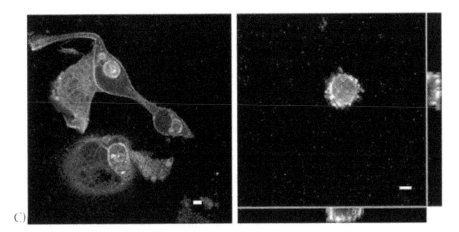

FIGURE 9   The CLSM images of HT-29 cells with incorporated PMN-II loaded with DOX. Cells were incubated with 50 µg of PMN-II loaded with DOX for 1 day (A), 3 days (B), and 6 days (C). HT-29 are labeled with FITC-Phalloidin (green fluorescence) while PMN-II were loaded with DOX (red fluorescence). Scale bars represent 5 µm.

17.4   CONCLUSION

The incorporation of NIPAAm residues in the PVA based network invokes a volume phase transition within a physiological temperature range. The CLSM and STXM allow the direct observation of shrinking behavior of thermoresponsive microgels in aqueous solution. The shrinking behavior upon temperature increase is monitored *insitu*. Moreover, NEXAFS studies confirm that the microgel particles chemical structure remained unchanged above the volume phase transition temperature.

The hydrophilic anticancer case drug DOX was loaded into the microgels. The release of DOX at 25°C (below LCST) was slower then at physiological temperature (above LCST), these results suggests that the DOX drug release kinetics strongly depend on environmental temperature, the swelling and the interactions of the loaded drug with microgels.

In vitro cytocompatibility tests allowed to assess that the interaction with these devices does not perturb the vitality of HT-29 cells and the interaction of the PMN-II microgels with HT-29 cells can be considered as potential injectable devices for further investigations on their impact on cell materials and tissues.

This finding will allow designing a microgel with implemented functionality by decorating the microsphere surface with hyaluronic acid, a ligand of CD44 receptor, the membrane protein overexpressed in tumor cells. The modification of the microgels surface will increase the bioavailability of DOX with a localized release of drug to the target cells.

KEYWORDS

- **Scanning transmission X-ray microscopy**
- **Confocal laser scanning microscopy**
- **Controlled drug delivery system**
- **Differential scanning calorimetry**
- **Microemulsion technique**

REFERENCES

1. Schmaljohann, D. Thermo and pH-responsive polymers in drug delivery. *Adv. Drug Del. Rev.*, **58**(15), 1655–1670 (2006).
2. Dimitrov, I., Trzebicka, B., Muller, A. H. E., Dworak, A., and Tsvetanov, C. B. Thermoresponsive water-soluble copolymers with doubly responsive reversibly interacting entities. *Prog. Polym. Sci.*, **32**(11), 1275–1343 (2007).
3. Zhang, X. Z. and Chu, C. C. Temperature-Sensitive Poly(*N*-Isopropylacrylamide)/Poly(Ethylen Glycol) Diacrylate Hydrogel Microspheres Formulation and Controlled Drug Release. *Am. J. Drug Deliv.*, **3**(1), 55–65 (2005).
4. Langer, R. and Peppas, N. A. Advances in Biomaterials, Drug Delivery, and Biotechnology. *AIChE J.*, **49**(12), 2990–3006 (2003).
5. Vakkalanka, S. K., Brazel, C. S., and Peppas, N. A. Temperature-and pH-sensitive terpolymers for modulated delivery of streptokinase. *J. Biomater. Sci. Polym. Ed.*, **8**(2), 119–129 (1996).
6. Zhang, X. Z., Zhou, R. X., and Cui, J. Z. A novel thermo-responsive drug delivery system with positive controlled release. *Int. J. Pharm.*, **235**(1-2), 43–50 (2002).
7. Hirokawa, Y. and Tanaka, T. Volume phase transition in a non-ionic gel. *J. Chem. Phys.*, **81**, 6379–6380 (1984).
8. Fujishige, S., Kubota, K., and Ando, I. Phase transition of aqueous solutions of poly(*N*-isopropylacrylamide) and poly(*N*-isopropylmethacrylamide). *J. Phys. Chem.*, **93**, 3311–3313 (1989).
9. Ghugare, S. V., Mozetic, P., and Paradossi, G. Temperature-Sensitive Poly(vinyl alcohol)/ Poly(methacrylate-co-*N*-isopropylacrylamide) Microgels for Doxorubicin Delivery. *Biomacromolecules*, **10**(6), 1589–1596 (2009).
10. Ghugare, S. V., Chiessi, E., Telling, M., Deriu, A., Gerelli, Y., Wuttke, J., and Paradossi, G. Structure and Dynamics of a Thermoresponsive Microgel around Its Volume Phase Transition Temperature. *J. Phys. Chem. B*, **114**(32), 10285–10293 (2010).
11. Ghugare, S. V., Chiessi, E., Fink, R., Gerelli, Y., Scotti, A., Deriu, A., Carrot, G., and Paradossi, G. Structural Investigation on Thermoresponsive PVA/Poly(Methacrylate-co-*N*-isopropylacrylamide) Microgels across the Volume Phase Transition. *Macromolecules*, **44**, 4470–4478 (2011).
12. Fitzgerald, P. A., Dupin, D., Armes, S. P., and Wanless, E. J. In Situ observations of adsorbed microgel particles. *Soft Matter*, **3**(5), 580–586 (2007).
13. Ade, H., Zhang, X., Cameron, S., Costello, C., Kirz, J., and Williams, S. Chemical contrast in X-ray microscopy and spatially resolved XANES spectroscopy of organic specimens. *Science*, **258**, 972–975 (1992).
14. Ade, H. and Hitchcock, A. P. NEXAFS microscopy and resonant scattering: Composition and orientation probed in real and reciprocal space. *Polymer*, **49**, 643–675 (2008).
15. Lawrence, J. R., Swerhone, G. D. W., Leppard, G. G., Araki, T., Zhang, X., West, M. M., and Hitchcock, A. P. Scanning Transmission Electron Microscopy Mapping of the Exopolymeric Matrix of Microbial Biofilms. *Appl. Environ. Microbiol.*, **69**, 5543–5554 (2003).
16. Toner, B., Fakra, S., Villalobos, M., Warwick, T., and Sposito, G. Spatially Resolved Characterization of Biogenic Manganese Oxide Production within a Bacterial Biofilm. *Appl. Environ. Microbiol.*, **71**, 1300–1310 (2005).

17. Koprinarov, I. N., Hitchcock, A. P., McCrory, C. T., and Childs, R. F. Quantitative Mapping of Structured Polymeric Systems Using Singular Value Decomposition Analysis of Soft X-ray Images. *J. Phys. Chem. B*, **106**(21), 5358–5364 (2001).

18. Mitchell, G. E., Wilson, L. R., Dineen, M. T., Urquhart, S. G., Hayes, F., Rightor, E. G., Hitchcock, A. P., and Ade, H. Quantitative Characterization of Microscopic Variations in the Cross-Link Density of Gels. *Macromolecules*, **35**(4), 1336–1341 (2002).

19. Harton, S. E., Lu¨ning, J., Betz, H., and Ade, H. Polystyrene/Poly(methyl methacrylate) Blends in the Presence of Cyclohexane: Selective Solvent Washing or Equilibrium Adsorption?. *Macromolecules*, **39**(22), 7729–7733 (2006).

20. De´jugnat, C., Ko¨hler, K., Dubois, M., Sukhorukov, G. B., Mo¨hwald, H., Zemb, T., and Guttmann, P. Membrane Densification of Heated Polyelectrolyte Multilayer Capsules Characterized by Soft X-ray Microscopy. *Adv. Mater.*, **19**(10), 1331–1336 (2007).

21. Tzvetkov, G., Graf, B., Fernandes, P., Fery, A., Cavalieri, F., Paradossi, G., and Fink, R. H. In situ characterization of gas-filled microballoons using soft X-ray microspectroscopy. *Soft Matter*, **4**, 510–514 (2008).

22. Cavalieri, F., Chiessi, E., Villa, R., Vigano,` L., Zaffaroni, N., Telling, M. F., and Paradossi G. Novel PVA-Based Hydrogel Microparticles for Doxorubicin Delivey. *Biomacromolecules*, **9**(7), 1967–1973 (2008).

23. Franssen, O. and Hennink, W. E. A novel preparation method for polymeric microparticles without the use of organic solvents. *Int. J. Pharm.*, **168**(1), 1–7 (1998).

24. De Belder, A. N. and Granath, K. Preparation and properties of fluorescein-labelled dextrans. *Carbohydr. Res.*, **30**(2), 375–378 (1973).

25. Raabe J., Tzvetkov, G., Flechsig, U., Böge, M., Jaggi, A., Sarafimov, B., Vernooij, M. G. C., Huthwelker, T., Ade, H., Kilcoyne, A. L. D., Tyliszczak, T., Fink, R. H., and Quitmann, C. PolLux a new facility for Soft X-Ray Spectromicroscopy at the SLS. *Rev. Sci. Instrum.*, **79**, 11370 (2008).

26. Winn, B., Ade, H., Buckley, C., Feser, M., Howells, M., Hulbert, S., Jacobsen, C., Kaznacheyev, K., Kirz, J., Osanna, A., Maser, J., McNulty, I., Miao, J., Oversluizen, T., Spector, S., Sullivan, B., Wang, Y., Wirick, S., and Zhang, H. Illumination for coherent soft X-ray applications: the new X1A beamline at the NSLS. *J. Synchrotron Radiat.*, **7**(6), 395–404 (2000).

27. Mosmann, T. Rapid colorimetric assay for cellular growth and survival: application to proliferation and cytotoxicity assays. *J. Immunol. Methods*, **65**(1-2), 55–63 (1983).

28. Fujji, S., Armes, S., Araki, T., and Ade, H. Direct Imaging and Spectroscopic Characterization of Stimulus-Responsive Microgels. *J. Am. Chem. Sco.*, **127**(48), 16808–16809 (2005).

29. Fujji, S., Dupin, D., Araki, T., Armes, S., and Ade, H. First Direct Imaging of Electrolyte-induced deswelling Behavior of pH-Responsive Microgels in Aqueous Media Using Scanning Transmission X-ray Microscopy. *Langmuir*, **25**(5), 2588–2592 (2009).

Index